Postharvest Management of Horticultural Produce

Recent Trends

The Authors

Dr R. T. Patil is former Director of Central Institute of Post Harvest Engineering and Technology (CIPHET) under ICAR which is nodal institute to undertake lead researches in Post Harvest Management and Value Addition.

Dr Desh Beer Singh is presently Director at ICAR-Central Institute of Temperate Horticulture, Srinagar, J&K. He did his M.Sc. (Horticulture) from Punjab Agricultural University, Ludhiana, Punjab and Ph.D. (Horticulture) from Bhim Rao Ambedkar University, Agra.

Dr R. K. Gupta former Director of ICAR – Central Institute of Post-Harvest Engineering & Technology (CIPHET), Ludhiana (Punjab) India.

Postharvest Management of Horticultural Produce

Recent Trends

R.T. Patil
Desh Beer Singh
R.K. Gupta
Central Institute of Postharvest Engineering and Technology
Malout Road, Abohar – 152 116, Punjab, India

2024
Daya Publishing House®
A Division of
Astral International Pvt. Ltd.
New Delhi - 110 002

First Published, 2009
Reprinted, 2024

ISBN: 9788170356080 (Hardbound)

Publisher's note:

Every possible effort has been made to ensure that the information contained in this book is accurate at the time of going to press, and the publisher and author cannot accept responsibility for any errors or omissions, however caused. No responsibility for loss or damage occasioned to any person acting, or refraining from action, as a result of the material in this publication can be accepted by the editor, the publisher or the author. The Publisher is not associated with any product or vendor mentioned in the book. The contents of this work are intended to further general scientific research, understanding and discussion only. Readers should consult with a specialist where appropriate.

Every effort has been made to trace the owners of copyright material used in this book, if any. The author and the publisher will be grateful for any omission brought to their notice for acknowledgement in the future editions of the book.

Published by : **Daya Publishing House®**
A Division of
Astral International Pvt. Ltd.
– ISO 9001:2008 Certified Company –
4736/23, Ansari Road, Darya Ganj
New Delhi-110 002
Ph. 011-43549197, 23278134
E-mail: info@astralint.com
Website: www.astralint.com

Digitally Printed at : **Replika Press Pvt. Ltd.**

Preface

Today's most urgent need is to increase the production of quality and nutritious food so that we may adequately feed the hungry/needy people. A major and often neglected step toward offering a greater volume of nutritious food is to prevent losses during postharvest handling of horticultural produce, which has been estimated as 30-40 per cent. With use of postharvest technologies to prevent their deterioration after harvest, supplies of fresh fruit and vegetables can be increased and postharvest losses can be minimized.

Fresh vegetables and fruits are vital source of vitamins, minerals and dietary fibers. Both provide essential ingredients like vitamins, minerals, carbohydrates, dietery fiber and supply complex carbohydrates and proteins. Green and yellow fruits and vegetables are rich source of vitamin A (β-carotene), thiamine, niacine, and folic acid which are required for normal functioning of human body are also present in significant quantity. Fruits and vegetables contribute over 90 per cent of dietary vitamin C. Because fruits and vegetables are perishable products with high metabolic activity during postharvest period, proper postharvest handling plays an important role in increasing their availability.

Remarkable improvement has been made in postharvest handling of various fruits and vegetables and in controlling postharvest diseases and disorders. Improved storage facilities have been developed for various fruits and vegetables that maintain the quality intact and adds to the common appeal. New chemicals and postharvest treatment techniques have been developed that are made effective in control of decay and attack of microbes. Proper harvesting, sorting, pre cooling and cold storage facilities are keeping pace with the needs of the industry.

Information on all these aspects is available in many journals, research papers, reviews, books, bulletins but are in scattered form. These informations in a documented form in the shape of a book, compiled in postharvest handling/management from harvesting to consumption fill the need throughout the country. We hope this publication will be useful to students of horticulture, marketing, food processing, engineers, nutritionists as well as processors and transporters engaged in this business.

R.T. Patil
Desh Beer Singh
R.K. Gupta

Contents

VEGETABLES

Introduction

Fruits and vegetables and their products not only improve the quality of the diet but also provide essential ingredients like vitamins, minerals, carbohydrates, dietary fibre etc. They also exhibit a high potential for generating employment in rural areas. Because of high productivity and value they provide and much better economic returns per unit areas compared to local crops, are a good source of foreign exchange.

India is the fruit and vegetable basket of the world. India being a home of wide variety of fruits and vegetables holds a unique position in production figures among other countries. Endowed with diverse agro climatic conditions such as tropical, sub tropical, arid and temperate, India holds a unique position for growing wide range of fruits, vegetables, spices and ornamental and floricultural crops plants like mango, citrus, banana, guava, papaya, grapes, onion, potato, brinjal, cauliflower, okra, tomato, orchids, roses, anthurium, spices like cumin, fenugreek, black pepper, cinnamon, cardamom, clove etc. In fact there is hardly any fruit or vegetable, which cannot be grown in India. India produces about 127560 Th. MT of fruits and vegetables accounts 9.22 per cent of production of world.

In addition to second largest producer of fruits and vegetables, India has emerged as the largest producer of coconut, arecanut, cashew nut, ginger, turmeric, black pepper and among the new crops, kiwi, olive crops and oil palm have been introduced with great success.

The changing scenario encourages private investment to go for hi-tech horticulture with micro-propagation, protected cultivation, drip irrigation, fertigation, and integrated nutrients and pest management, besides making use of latest postharvest measures, particularly in the case of perishable commodities. As a result, horticulture crop production has moved from rural confines to commercial ventures.

After a number of revolutions since 1965- the green revolution (cereals), the white revolution (milk), blue revolution (fish) and yellow revolution (oil seeds)-it is the time now for a "Golden revolution" in horticultural sector. However their perishable nature is a major bottleneck in realizing their remunerative price to the farmers. The perishability leads to losses as high as 25-40 per cent, high cost for consumers, however distress sale of the growers. Hence, proper postharvest management is key to golden revolution in horticulture sector.

The covered area under the fresh fruits in India was 5510000 ha with the production of 58740000 MT in 2005-06 accounts 10 per cent of world production. A large variety of fruits are grown in India, of those, mango, banana, citrus, pineapple, papaya, guava, sapota, jackfruit, litchi, and grape, among the tropical and sub-tropical fruits; apple, pear, peach, plum, apricot, almond and walnut among the temperate fruits; and aonla, ber, pomegranate, anona, fig, phalsa among the arid zone fruits are important. It leads the world in the production of mango, banana, sapota and acid lime and has recorded highest productivity in grape. Mango is the most important fruit covering about 39 per cent of the area and accounts for 23 per cent of total fruit production in the country. India's share in world production of mango is about 54 per cent. Citrus ranks second in area and accounts for about 10 per cent of total fruits in the country. Limes, lemons, sweet orange and mandarin, cover bulk of the area under this group of fruits. Banana ranks third in area covering about 13 per cent of the total area. India occupies the first position in banana production. Fruits like guava, papaya is about four per cent and litchi is about one per cent. The arid zones of the country are potential areas for fruits like aonla, ber, pomegranate, and anona. There has been a steady increase in the area and production of these fruits particularly aonla, ber and pomegranate in the country as a result of identification and development of suitable varieties and production technologies. In addition to these, date palm and fig cultivation are also finding favour in suitable areas. Mango, called the king of fruits in India, accounts for 40 per cent of the national fruit production of 22.168 million tonnes a year. It occupies 42 per cent of the country's 24.87 million hectares land under fruit cultivation. Over 90 per cent of India's exports in fresh products goes to west Asia and East European markets. However, it needs to augment its food and processing industry at a mega scale. This is due to its potential in different agro climatic zones. India's export of fresh fruits has increased from Rs. 225.67 crores in 2005-06 to 256.43 crores in 2006-07. India's export of fresh fruit and vegetable have increased from Rs. 1658.72 crores in 2005 06 to Rs. 2411.66 crores in 2006-07. India is the second largest producer of vegetables that account about 16 per cent of the world production. The area under vegetable was covered 7164000 ha with a production of 109050 MT in 2005-06. India's export of other vegetable has increased from Rs. 267.69 crores in 2005-06 to Rs. 430.02 crores in 2006-07. Although India is one of the largest producer of fruits in the world, the production per capita is only about 100 grams per day. However, it is estimated that more than 20-22 per cent of the total production of fruits is lost due to spoilage at various postharvest stages. Thus, the per capita availability of fruits is further reduced to around 80 grams per day which is almost half the requirement for a balanced diet. The fruit production in India has recorded a growth rate of 3.9 per cent, whereas the fruit-processing sector has grown at about 20 per cent per annum. However, the growth rates have been extensively higher for frozen fruits and vegetables (121 per cent) and dehydrated fruits and vegetables (24 per cent). Over 4000 fruit processing units exists in India with an aggregate capacity of more than 12 lakh MT (less than 4 per cent of total fruits produced).

More than 40 kinds of vegetables belonging to different groups, namely, solanaceous, cucurbitaceous, leguminous, cruciferous (cole crops), root crops and leafy vegetables are grown in India in tropical, subtropical and temperature regions. Important vegetable crops grown in the country are tomato, onion, brinjal, cabbage, cauliflower, okra and peas. India is next only to China in area and production of vegetables. India contributes about 13 per cent to the world vegetable production and occupies first position in the production of cauliflower, second in onion and third in cabbage in world.

Abundant investment opportunities are there in expanding the export market. An increasing acceptance of new products with market development efforts has been witnessed lately given the fact that there is a good international demand for certain fruits and vegetable products. India ranks fifth in the world in cropped area under cultivation and production of potatoes. India produces 41 per cent of world's mangoes, 23 per cent bananas, 24 per cent cashew nuts, 36 per cent green peas and 10 per cent onion. The total export value of the main exporting fruit crop from India is mango. Exports of mangoes, grapes, mushrooms have started going to the United Kingdom, Middle East, Singapore and Hong Kong, and among vegetable, onion occupies first position potatoes and green vegetables like okra, bitter gourd, green chillies have good export potential.

Table 1: Production of Major Fruits in India (2005-06)

Crop	*Production 000MT*
Mango	10817
Citrus	4760
Pineapple	1300
Banana	16820
Pomegranate	500
Papaya	7000

Table 2: Production of Major Vegetables in India (2005-06)

Crop	*Production 000MT*
Brinjal	8200
Tomato	7600
Cauliflower	4800
Cabbage	6000
Potato	25000

Present Status of Postharvest and Processing

The fresh fruits are mostly harvested by hand or hand tools. Sorting and grading of fruits are done on a very limited scale and that too are based on visual inspection only. Limited precooling facilities are available for highly perishable produce like grapes; strawberries etc. that too only for export purpose. It is estimated that around 20-25 per cent of the total vegetables is lost due to poor postharvesting practices. Less than 2 per cent of the total vegetables produced in the country are commercially processed as compared to 70 per cent in Brazil and 65 per cent in USA. Around 1.5 lakh MT of vegetable is sold as processed products. Postharvest handling and processing plays an important role in the conservation and effective utilizations of fruits and vegetables. The processing also helps in generating rural employment; besides, processed fruits and vegetables are a source of earning foreign exchange. India has favorable climatic conditions and vast potential for growing fruits and vegetables. However, slightly over one per cent of the total production of fruits and vegetables is processed. This is because of a number of problems faced by the industry such as high cost of production, processing, packaging and transportation. The prominent processed items are ready to serve beverages, fruit pulps and juices, fruit based ready-to-serve beverages, canned fruits and vegetables, jams, squashes, pickles, chutneys and dehydrated vegetables. More recently, products like frozen fruit pulps dehydrated and freeze dried vegetables are also available in processing sector. Frozen fruits and vegetables products, fruits juice concentrates and vegetable curry in pouches, canned mushroom and mushroom products have been taken up by the fruits and vegetables processing industries.

Postharvest Management and Magnitude of Losses

The most important consideration for the present is increasing the production of nutritious food so that we can adequately feed the hungry people on the planet. A major and after neglected step towards offering a great volume of the nutritious food is to prevent losses between harvesting and consumption. Fruits and vegetables are highly perishable commodities and due to lack of adequate postharvest handling facilities and proper infra structure in our country, the postharvest losses in our country are very high. Actual postharvest losses have been estimated to be as high as 25-30 per cent of the produce. India's wastage of fruits and vegetables are estimated to about Rs. 33000 crores in value terms due to absence of proper postharvest handling and storage facilities. Even if 10 per cent of fruits could be saved by counting them into processed products at the peak production, there will be a saving of Rs. 650 crores to the horticultural wealth of the country. But the fruits and vegetable processing industry at present is utilizing only about 1.8 per cent (Table 4) of total production for processing as against the performance of other countries like Malayasia (83 per cent), Phillipines (78.0 per cent), Brazil (70 per cent) and Arabia (83 per cent).

Table 3: Production of Different Types of Horticultural Commodities and their Estimates of Postharvest Losses

Sl.No.	Commodity	Present Level of Production		Postharvest Losses		
		Quantity (Mt)*	*Average Price (Rs/t)*	*%***	*Quantity (Mt)*	*Monetary Value (Rs. in Crore)*
1.	Fruits (Mango, Citrus Pineapple, Bananas, Pomegranate, Papaya etc.)	58740000	16000	35	20559000	32894.40
2.	Vegetables (Brinjal, Tomato, Cauliflower, Cabbage, Potato etc.)	109050	15000	35	38167.50	57.25
					Total	32951.65

*: www.apeda.com; **: www.commodityonline.com

Table 4: Horticultural Commodity Used for Processing (International Status)

Country	*% Fruits and Vegetables Used for Processing*
Brazil	70
Phillipines	78
S.Arabia	80
Malayasia	83
India	1.8
UK	50
Australia	60

Fruits

1

Grape

Botanical Name: *Vitis vinifera*

Family: Ampelidaceae (Vitaceae)

1. Introduction

Grapes are the most widely cultivated fruit crop in the world. Among the continents, Europe is the largest producer of grapes. Grapes are grown mostly for wine making in Italy, France, Spain, the United States, Turkey, Argentina, and South Africa. Evidence shows that invaders from Afghanistan and Persia introduced grapes in to India in AD 1300. Then it spread to grape growing areas of India such as Maharashtra, Karnataka, Andhra Pradesh and Tamil Nadu and later to North India.

2. Composition (Per 100g Edible Portion)

The chemical composition of grapes varies according to variety and the environment under which the grapes are grown. General composition of grapes includes:

Constituent	*Composition*	
	Blue Variety	*Pale Green Varieties*
Moisture (g)	82.2	79.2
Protein (g)	0.6	0.5
Fat (g)	0.4	0.3
Minerals (g)	0.9	0.6
Fibre (g)	2.8	2.9
Carbohydrates (g)	13.1	16.5
Energy (Kcal)	58	71
Calcium (mg)	20	20
Phosphorus (mg)	23	30
Iron (mg)	0.50	0.52
Carotene (μg)	3	0
Thiamine (mg)	0.04	0
Riboflavin (mg)	0.03	0
Niacin (mg)	0.2	0
Vitamin C (mg)	1	1

3. Suitable Cultivars

Table Varieties

Almeria, Calmeria, Cardinal, Perlette, Thompson Seedless, Beauty Seedless, Perlette, Pusa Seedless, Delight, Cordina, Bangalore Blue.

Raisin Varieties

Thompson Seedless, Muscat of Alexandria, Black Corinth, Seedless Sultana

Wine Varieties

Red wine: Alealico, Alicanle Barbera, Cabernet Sauvignon, Missim, Refosco, Ruby Cabernel, Salvador, Touriga, Zinfandel

White wine: Aligole, Burger, Chardonnay, Folte Blanche, French Colombard, Gray, Riesling, Muscat Blanc, Palomino, Pinol Blanc, Semillon, White Reisling

Juice Grapes

Beauty Seedless, Early Muscat, Champion, Black Champa, and Bangalore Blue.

4. Maturity Indices

Wine grapes are picked by hand or mechanical harvesters. Table grapes are harvested based on the texture of the pulp, peel, color, easy separation of the berries from the bunches and characteristic aroma. Besides these, 12-14 per cent TSS for Anab-e-Shahi, and 19-20 per cent TSS for Thompson Seedless and Selection-7 are used as harvesting index. Grapes for dry wine should have high acidity and moderate sugar content. Therefore, such grapes are usually harvested at 20-24° Brix.

5. Quality Indices

High consumer acceptance is attained in the fruit when high SSC or SSC/TA ratio is reached. Berry firmness is also an important factor for consumer acceptance. The fruit should be free of defects such as decay, cracked berries, stem browning, shriveling, sunburned or dried berries, and insect damage.

6. Postharvest Management

Harvesting

Grapes should be harvested during cool time of the day. The grape bunchs are trimmed/cut with trimmers. Grapes harvested at firm-ripe stage transported and stored better than the under ripe or over ripe stages. Over ripe berries are prone to rapid attack of decay organisms, and in some varieties they tend to shatter rapidly.

Precooling

Precooling in the general rule means that within the limits at which fresh grapes are usually handled. Usually a reduction in temperature of 9.5°C halves the rate of respiration and doubles the shelf life. A fruit temperature below 4.4°C greatly retards the development of all fungi and prohibits their growth. Precooling thus checks stemrotting, browning, and berry shattering

Grading

Grading of grapes can be done according to size, sugar content, and appearance. After grading of grapes, "culled" produce is separated and can be used for other purposes such as wine making, feed for cattle, or for making any other by-product.

Packaging

Corrugated fiber board (CFB) cartons of 2-4 kg capacity are recommended for packaging of grapes. The application of 0.75 per cent calcium nitrate as a preharvest spray on Perlette grapes 10 days before harvest reduce weight loss, berry drop, and decay. Application of cycocel and Alar (2000-4000 ppm) and kinetin (50-150 ppm) used as preharvest spray reduced berry rot and berry shatter during storage. Planofix at 100-150 ppm as a preharvest spray 2 weeks prior to harvest reduces berry drop. Cycocel and Alar at 100 ppm treated at harvest stage reduce storage losses from 8.5 per cent to 4.2 per cent.

Storage

Grapes may be cold-stored to prolong shelf life and to relieve seasonal gluts by extending their market period. The most important factors influencing the storage life of grapes are temperature, relative humidity, air movement, and fumigation with SO_2.

An evaporative cooled (EC) storage structure at CIPHET, Ludhiana can be efficiently used for on farm storage.

(*a*) Optimum Temperature

For better quantity retention berry storage at -1.0 to 0° C is recommended. The highest freezing point for berries is -2.1° C, but freezing point varies depending on Soluble Solid Content.

(*b*) Optimum Relative Humidity

90-95 per cent RH and an air velocity of approximately 20-40 feet per minute (FPM) is suggested during storage.

(*c*) Rates of Respiration

Temperature	*ml CO_2/kg hr**
0° C	1-2
5° C	3-4
10° C	5-8
20° C	12-15

Stem respiration rate is approximately 15 times higher than berry respiration.

* To calculate heat production, multiply ml CO_2/kg hr by 122 to get kcal/metric ton/day.

(*d*) Rates of Ethylene Production

<0.1 µl/kg·hr at 20° C

(*e*) Responses to Ethylene

Table grapes are not very sensitive to ethylene. However, exposure to ethylene (>10 ppm) may be a secondary factor in shatter.

(*f*) Responses to Controlled Atmospheres (CA)

CA (2-5 per cent O_2 + 1-5 per cent CO_2) during storage/shipment is not recommended for table grapes because its benefit is very little and SO_2 is already used for decay control.

(*g*) Irradiation

Irradiation of Thompson Seedless grapes with 1 and 2 x 10^5 rad and stored at 4.4°C are well liked and remains in marketable condition up to 1 month.

SO_2 Injury

7. Postharvest Disorders, Diseases and Their Control

Physiological Disorders

Blossom End Rot

A black sunken spot develops at the blossom end of the berry, which later on spreads with water soaked region around it. Defective calcium nutrition and assimilation is the main cause for it. Spray of 1.0 per cent calcium nitrate can correct this disorder.

Shatter (Loss of Berries from the Cap Stem)

In general, shatter increases in severity with increasing maturity, *i.e.*, the longer the fruit remains on the vine. Shatter varies considerably from season to season, and there is a large difference among varieties. Gibberellin applied at fruit set weakens berry attachment. Shatter occurs mainly due to rough handling during field packing with additional shatter occurring all the way to the final retail sale. Shatter incidence can be reduced by controlling pack depth and fruit packing density using cluster bagging, gentle handling and maintaining recommended temperature and relative humidity

Waterberry

It is associated with fruit ripening and most often begins to develop shortly after veraison (berry softening). The earliest symptom is the development of small (1-2 mm) dark spots on the cap stems (pedicles) and/or other parts of the cluster framework. These spots become necrotic, slightly sunken, and expand to affect more areas. The affected berries become watery, soft, and flabby when ripe. Avoid over fertilization with nitrogen. Foliar nutrient sprays of nitrogen should be avoided in waterberry-prone vineyards. Trimming off affected berries during harvest and packing is a common practice.

Pathological Disorders

Gray Rot

The most common fungus to attack the fruit of grapes is *Botrytis cinerea,* which causes gray rot. Damage can be severe if there is a simultaneous attack by *Penicillium* or *Aspergillus.* Gray mold is the most serious disease of grapes during storage at low temperature, and it originates from late-season infections in the vineyard. Gray mold is the most destructive of the postharvest diseases of table grapes, primarily because it develops at temperatures as low as -0.5° C and grows from berry to berry. Gray mold first turns berries brown, then loosens the skin of the berry, its white, thread-like hyphal filaments erupt through the berry surface, and finally masses of gray colored spores develop. Wounds near harvest also provide opportunities for infections. No wound is required for infection when wet conditions occur. Botrytis infection can be reduced by removing desiccated, infected grapes of the previous season from vines,

Grey Mould

leaf-removal canopy management, preharvest fungicides, trimming visibly infected, split, cracked, or otherwise damaged grapes before packing, immediate cooling and fumigation with sulfur dioxide (100 ppm for one hour).

2
Mango

Botanical Name: *Mangifera indica*

Family: Anacardiaceae

1. Introduction

Mango is known to have originated in Southeast Asia. The natural spread of the genus is limited to the Indo-Malaysian region. The genus is reported to contain 41 species, but only *Mangifera indica* has been cultivated and includes almost all the edible cultivars. Several hundred varieties exist in India, but only a few specific cultivars are commercialized. India contributes about 64 per cent of the world's production. Other prominent mango-producing countries are Mexico, Pakistan, Brazil, the Philippines, and Thailand.

2. Composition (Per 100g edible portion)

The composition of mango varies with location of cultivation, variety, and stage of maturity.

The general composition includes:

Constituent	*Composition*	*Constituent*	*Composition*
Moisture (%)	81	Carotene (μg)	2743
Protein (g)	0.6	Thiamine (mg)	0.08
Fat (g)	0.4	Riboflavin (mg)	0.09
Minerals (g)	0.4	Niacin (mg)	0.9
Fiber (g)	0.7	Vitamin C (mg)	16
Carbohydrates (g)	16.9		
Energy (Kcal)	74		
Calcium (mg)	14		
Phosphorus (mg)	16		
Iron (mg)	1.3		

Chemical Composition of Ripe Mangoes of Some Important Mango Varieties of India

Variety	*Per cent Fresh Weight (Pulp)*					
	Total Soluble Solids (Brix)	*Acidity (Malic Acid) (%)*	*Alcohol-insoluble Residue (%)*	*Total Sugars (%)*	*Reducing Sugars (%)*	*Vitamin C (mg/100g))*
Alphonso	17–20	0.14–0.64	1.0–2.5	10.5–18.5	2.5–4.0	50–80
Baneshan	14–19	0.15–0.30	1.5–2.2	10.5–15.5	4.5–7.0	25–35
Pairi	14–16	0.10–0.34	1.0–3.0	11.6–15.6	2.5–5.2	10–25
Totapuri	14–16	0.20–0.45	1.0–2.5	11.2–15.4	4.0–5.8	10–20
Neelum	16–18	0.15–0.30	1.0–2.2	11.4–15.5	5.0–7.0	10–25
Mulgoa	14–20	0.10–0.25	2.0–3.5	15.0–15.5	3.2–4.0	20–30
Dashehari	18–22	0.20–0.30	1.5–2.2	13.5–16.0	2.5–4.0	25–50
Fazli	18–20	0.10–0.20	1.2–2.0	12.4–15.5	5.0–7.5	75–100
Langra	18–22	0.20–0.35	1.4–2.2	12.1–14.0	2.4–3.5	100–175
Chausa	18–24	0.20–0.35	1.4–2.4	16.0–18.0	2.0–3.0	30–50

3. Suitable Cultivars

The prominent varieties are Alphonso, Totapuri, Baneshan, Bombai, Bombay Green, Dashehari, Fajri/Fazli, Himsagar, Kesar, Kishan Bhog,

Langra, Mulgoa, Neelum, Chausa, Suvarnarekha, Vanraj, Zardalu, and Gulab Khas. Furthermore, two mango hybrids, Mallika (Neelum x Dashehari) and Amrapali (Dashehari x Neelum) are released from the Indian Agricultural Research Institute (IARI), New Delhi for their commercial potential.

4. Maturity Indices

The maturity of fruit has been correlated with various physical characteristics such as surface color, shape, size, shoulder growth and specific gravity, and chemical parameters such as soluble solids, titratable acidity, starch, phenolic compounds, and carotenoids.

- ✰ Change in fruit shape (fullness of the cheeks).
- ✰ Change in skin color from dark-green to light green to yellow (in some cultivars). Red color on the skin of some cultivars is not a dependable maturity index
- ✰ Change in flesh color from greenish-yellow to yellow to orange.

Non-destructive Methods for Determining Maturity of Fruits Developed at CIPHET, Ludhiana

Eating quality and postharvest shelf life of ripe fruits depends on its maturity at harvest. Change of peel color and total soluble solids (TSS) are indicative parameters to measure maturity of most of the fruit. A maturity index is defined based on TSS and colour values. The model was developed to predict maturity using colourimeter Colour and maturity index chart is also developed to determine maturity. This technique can be employed to sort the fruits at well-organized markets and in processing plant.

5. Quality Indices

- ✰ Uniformity of shape and size; skin color (depending on cultivar); flesh firmness.
- ✰ Freedom from decay and defects, including sunburn, sapburn, skin abrasions, stem-end cavity, hot water scald, chilling injury, and insect damage.
- ✰ Changes associated with ripening include starch to sugar conversion (increased sweetness), decreased acidity and increased carotenoids and aroma volatiles.

☆ There are large differences in flavor quality (sweetness, sourness, aroma) and textural quality (fiber content) among cultivars.

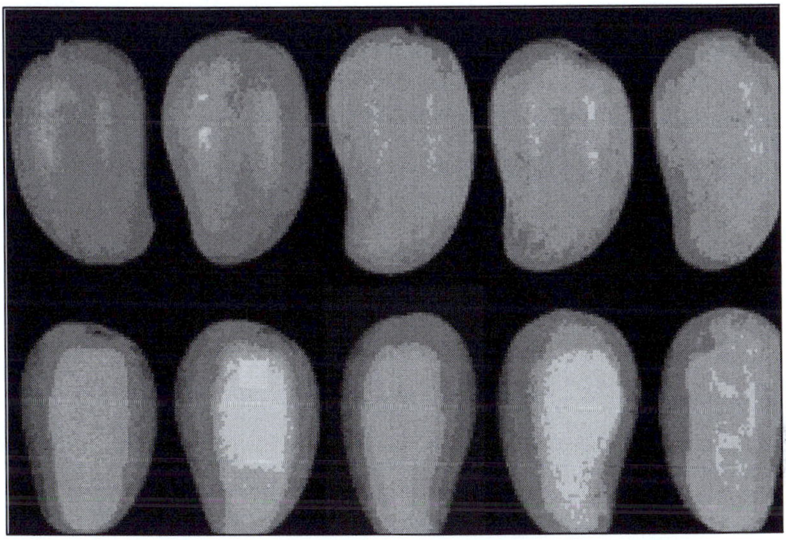

Fruit Ripening and Maturity

6. Postharvest Management

Harvesting

Generally fruits are hand picked or plucked with a harvester. Fruits not reached by hand are most often retrieved with poles adapted with a severing blade and a bag (Technology developed at AICRP PHT center at Bangalore).

Grading/Sorting

Grading in mango is being done on the basis of specific gravity, fruit weight or size.

Packing

Proper packing is an essential prerequisite. Mangoes packed in CFB boxes with partitions show less bruising, slow ripening, reduces shriveling and less spoilage.

Responses to Ethylene to Ripening

The response of mango fruit to postharvest ripening harmone (ethrel) treatment is not consistent throughout its life cycle, and its age at the time of harvest has a pronounced effect on ripening.

Exposure to 100 ppm ethylene for 12 to 24 hours at 20 to 22°C and 90-95 per cent relative humidity results in accelerated and more uniform ripening of mangoes within 5-9 days, depending on cultivar and maturity stage. Carbon dioxide concentration should be kept below 1 per cent in the ripening room.

Storage

For short term storage evaporative cooled storage structure developed at CIPHET, Ludhiana is useful for mango storage.

For transportation of mango to precooling chamber developed at CIPHET, Ludhiana/AICRP Centre can be effectively used.

(*a*) Optimum Temperature

13°C for mature-green mangoes

10°C for partially-ripe and ripe mangoes

(*b*) Optimum Relative Humidity

90-95 per cent

(*c*) Rates of Respiration/CO_2 Production

Temperature	10°C	13°C	15°C	20°C
ml CO_2/kg hr	12-16	15-22	19-28	35-80

To calculate heat production multiply ml CO_2/kg·hr by 122 to get kcal/metric ton/day.

(*d*) Rates of Ethylene Production

Temperature	10°C	13°C	15°C	20°C
µl C_2H_4/kg hr	0.1-0.5	0.2-1.0	0.3-4.0	0.5-8.0

(*e*) Responses to Controlled Atmospheres (CA)

- Optimum CA 3-5 per cent O_2 and 5-8 per cent CO_2
- CA delays ripening and reduces respiration and ethylene production rates. Postharvest life potential at 13°C: 2-4 weeks in air and 3-6 weeks in CA, depending on cultivar and maturity stage.
- Exposure to below 2 per cent O_2 and/or above 8 per cent CO_2 may induce skin discoloration, grayish flesh color, and off-flavor development.

7. Postharvest Diseases, Disorders and Control

Physiological and Physical Disorders

Sapburn

Dark-brown to black discoloration of mango skin due to chemical and Physiological injury from exudate (sap) from cut stem.

Skin Abrasions

Abrasions due to fruit rubbing against rough surfaces or each other result in skin discoloration and accelerated water loss.

Cold Storage/Low Temperature Injury

Symptoms include uneven ripening, poor color and flavor, surface pitting, grayish scald-like skin discoloration, increased susceptibility to decay, and, in severe cases, flesh browning. Chilling injury incidence and severity depend on cultivar, ripeness stage (riper mangoes are less susceptible) and temperature and duration of exposure.

Heat Injury

Exposure to temperatures above 30°C for longer than 10 days results in uneven ripening, mottled skin and strong flavor. Exceeding the time and/or temperature combinations recommended for decay and/or insect control, such as 46.4°C water dip for 65-90 minutes (depending on fruit size) causes heat injury (skin scald, blotchy coloration, uneven ripening).

Internal Flesh Breakdown (Stem-end Cavity)

Flesh breakdown and development of internal cavities between seed and peduncle. This disorder is more prevalent in tree-ripened mangoes.

Jelly-seed (Premature Ripening)

Disintegration of flesh around seed into a jelly-like mass.

Soft-Nose

Softening of tissue at apex. Flesh appears over-ripe and may discolor and become spongy. This disorder may be related to calcium deficiency.

Pathological Disorders

Anthracnose

Caused by *Colletotrichum gloesporioides*, begins as latent Disorders infections in unripe fruit and develops when the mangoes begin to ripen.

Lesions may remain limited to the skin or may invade and darken the flesh.

Diplodia Stem-end Rot

Caused by *Lasiodiplodia theobromae*, affects mechanically-injured areas on the stem or skin. The fungus grows from the pedicel into a circular black lesion around the pedicel.

Control Strategies

1. Careful handling to minimize mechanical injuries.
2. Hot water treatment: 5-10 minutes (depending on fruit size) dip in 50°C ± 2°C water.
3. Postharvest fungicide (thiabendazole) treatment alone or in combination with hot water treatment maintaining optimum temperature and relative humidity during all handling steps.

Internal Breakdown

Generally, the symptoms of internal breakdown are characterized by breakdown of the flesh on the ventral side and toward the apex in the fruit. At the advanced stage of the disorder, the tissue becomes spongy and grayish black. The disorder commences from the stone and spreads

Cold Storage/Low Temperature Injury

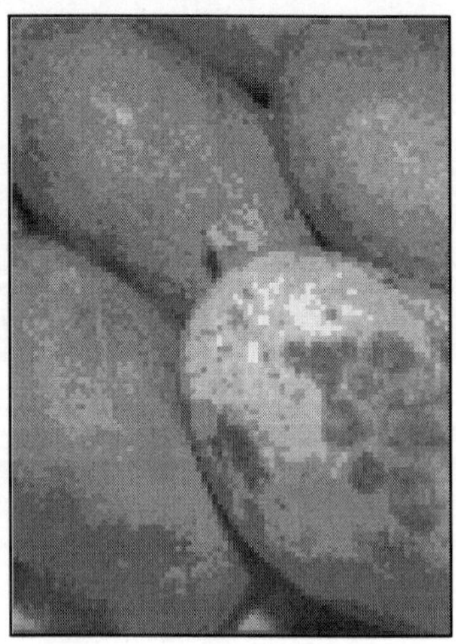

Anthracnose

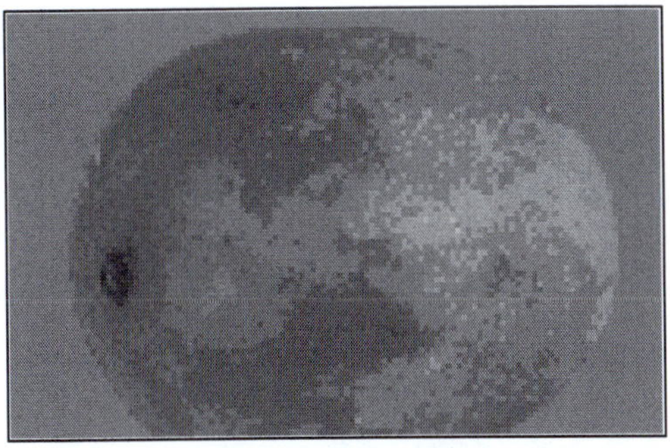

Stem End Rot Jelly

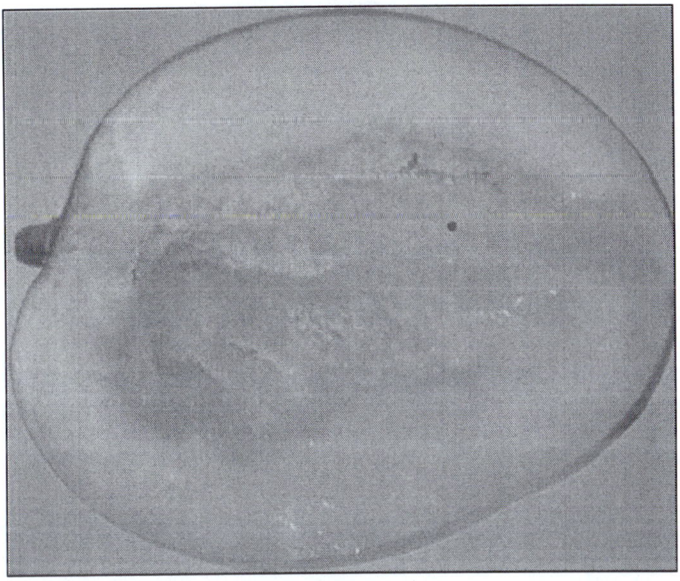

Jelly Seed

toward the periphery. In severe cases, the whole fleshy tissue becomes too soft, resembling bacterial rot. A survey report suggested that 25-30 per cent of the Alphonso crop could be affected by this disorder.

3

Pineapple

Botanical Name: ***Ananas comosus***

Family: Bromeliaceae

1. Introduction

The pineapple is one of the most important commercial fruits of the world. It is widely cultivated throughout the tropics and sub tropics. It is an important economic fruit crop in Thailand, Philippines, China, Brazil, Hawaii, India, Mexico, and South Africa. In India it is cultivated in Tripura, Assam, Meghalaya, Manipur, Kerala, Karnataka, West Bengal. Most commercial production is used in processing. Pineapples are used as dessert fruits or for preparation of canned pineapple in the form of slices or rings, and in the preparation of juices and jams. The fruits are commonly used in fruit salad along with banana and papaya. Alcohol, calcium nitrate, citric acid, and vinegar can be prepared from pineapple. The osmodehydrated pineapple candy is also a commercially possible product.

2. Composition (Per 100g of Edible Portion)

Constituent	*Composition*	*Constituent*	*Composition*
Water (per cent)	81.2–86.2	Vitamin B6 (per cent)	10–140
Brix (per cent)	10.8–17.5	Pigments (ppm of carotene)	0.2–2.5
Sucrose (per cent)	5.9–12.0	Carotene (mg)	0.13–0.29
Glucose (per cent)	1.0–3.2	Xanthophylls (mg)	0.03
Fructose (per cent)	0.6–2.3	Esters (ppm)	0.2–2.5
Cellulose (per cent)	0.43–0.54	Vitamins (μ/100g) fresh wt	
Pectin (per cent)	0.06–0.16	Aminobenzoic acid	17–22
Titrable acid (per cent)	0.6–1.62	Folic acid	2.5–4.8
Citric acid (per cent)	0.32–1.22	Niacin	200–280
Malic acid (per cent)	0.1–0.47	Pantothenic acid	75–163
Oxalic acid (per cent)	0.005	Thiamine	69–125
Ash (per cent)	0.30–0.42	Riboflavin	20–88
Fiber (per cent)	0.30–0.61	Vitamin A	0.02–0.04
Nitrogen (per cent)	0.045–0.115	Ascorbic acid	10–25
Ether extract (per cent)	0.2		

3. Suitable Cultivars

There is no reliable botanical or horticultural classification of pineapple varieties. The important commercial varieties grown in leading pineapple-growing countries are as follows:

Country	*Varieties*
Hawaii	Cayenne and Hillo
Phillipines	Cayenne
Malaysia	Singapore Spanish
Australia	Cayenne and Red Spanish
Kenya	Cayenne
Cuba	Cayenne and Red Spanish
Taiwan	Cayenne
Brazil	Red Spanish and Abacaxi
India	Giant Kew, Queen, Maritius

4. Maturity Indices

Change of shell color from green to yellow at the base of the fruit. Pineapples are non-climateric fruits and should be harvested when ready to eat. A minimum soluble solids content of 12 per cent and a maximum acidity of 1 per cent will assure minimum flavor acceptability by most consumers.

1 2 3 4 5 6 7

Different Stages of Maturity

5. Quality Indices

Uniformity of size and shape; firmness; free from decay, sunburn, sunscald, cracks, bruising, internal breakdown, endogenous brown spot, gummosis, and insect damage.

Tops (crown leaves): green color, medium length, and straightness.

Range of soluble solids = 11-18 per cent; titratable acidity (mainly citric acid) = 0.5-1.6 per cent; and ascorbic acid = 20-65 mg/100g fresh weight, depending on cultivar and ripeness stage.

6. Postharvest Management

Harvesting

Harvesting is mostly done by hand. Harvested fruits are trimmed (Crown removed and bases cleaned) and placed in wooden boxes, loaded on flat bed trucks. Pineapple fruits require great care while handling, and any injury to the fruit leads to fruit rot. Packing is necessaryfor transport to distant places. Fruits are wrapped in paper or paddy straw and arranged in bamboo baskets.

Storage

(*a*) Optimum Temperature

10-13° for partially-ripe pineapples

7-10°C for ripe pineapples.

(*b*) Optimum Relative Humidity

85-90 per cent

(*c*) Rates of Respiration/CO_2 Production

Temperature	7°C	10°C	13°C	15°C
ml CO_2/kg hr	2-4	3-5	5-8	8-10

To calculate heat production multiply ml CO_2/kg·hr by 122 to get kcal/metric ton/day.

(*d*) Rates of Ethylene Production

Less than 0.2 μlC_2H_4/kg·hr at 20°C

(*e*) Responses to Ethylene

Exposure of pineapples to ethylene may result in slightly faster degreening (loss of chlorophyll) without influencing internal quality. Pineapples must be picked when ripe because they do not continue to ripen after harvest.

(*f*) Responses to Controlled Atmospheres (CA)

- ☆ 3-5 per cent O_2 and 5-8 per cent CO_2
- ☆ Benefits of CA include delayed senescence and reduced respiration rate
- ☆ Postharvest life potential: 2-4 weeks in air and 4-6 weeks in CA 10°C, depending on cultivar and ripeness stage
- ☆ Exposure to O_2 levels below 2 per cent and/or CO_2 levels above 10 per cent should be avoided because of the potential for development of off-flavors.
- ☆ Waxing may be used to modify O_2 and CO_2 concentrations within the fruit enough to reduce incidence and severity of endogenous brown spot.

7. Postharvest Diseases and Disorders

Physiological and Physical Disorders

Chilling Injury

Exposure of pineapples to temperatures below 7°C results in chilling injury. Ripe fruits are less susceptible than unripe or partially-ripe fruits. Symptoms include dull green color when ripened (failure to ripen properly), water- soaked flesh, darkening of the core tissue, increased susceptibility to decay, and wilting and discoloration of crown leaves.

Black Heart

Is usually associated with exposure of pineapples before or after harvest to chilling temperatures, *e.g.* below 7°C for one week or longer. Symptoms are water-soaked, brown areas that begin as spots in the core area and enlarge to make the entire center brown in severe cases. Waxing is effective in reducing chilling injury symptoms. A heat treatment at 35°C for one day has been shown to ameliorate EBS symptoms in pineapples transported at 7°C by inhibiting activity of polyphenol oxidase and consequently tissue browning.

Pathological Disorders

Thielaviopsis Rot (Black Rot, Water Blister)

Caused by *Thielaviopsis paradoxa,* may start at the stem and advance through most of the flesh with the only external symptom being slight

Brown Spot

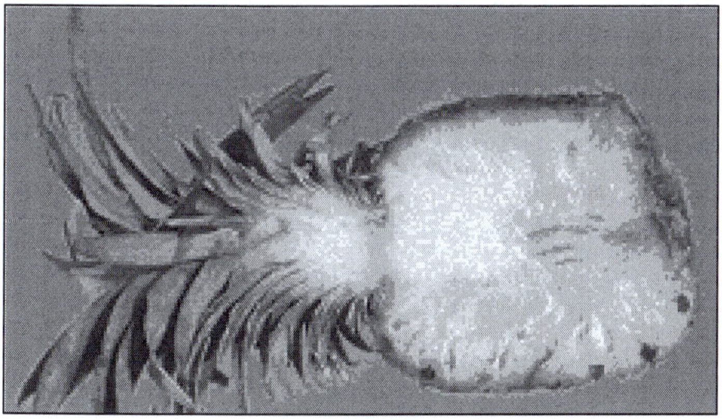

Water Blister

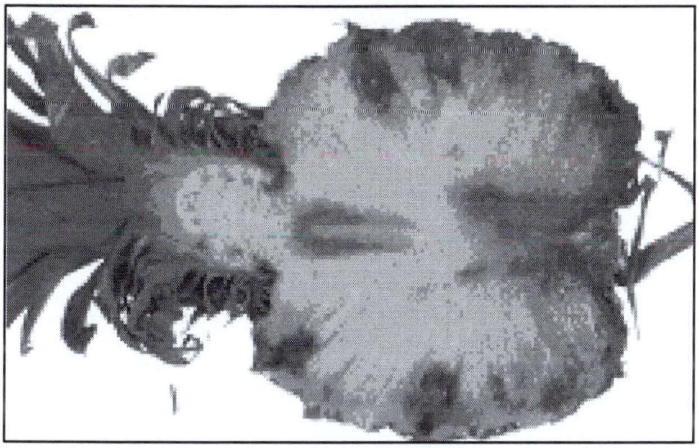

Fruitlet Core Rot

skin darkening due to watersoaking of the skin over rotted portions of the flesh. As the flesh softens, the skin above readily breaks under slight pressure.

Yeast Fermentation

Caused by *Saccharomyces* spp., is usually associated with overripe fruit. The yeast enters the fruit through wounds. Fruit flesh becomes soft and bright yellow and is ruptured by large gas cavities.

Control Strategies

- ✰ Careful handling to minimize mechanical injuries
- ✰ Prompt cooling and maintenance of optimum temperature and relative humidity throughout postharvest handling operations.
- ✰ Application of fungicides, such as thiabendazole (TBZ).

4

Papaya

Botanical Name: ***Carica papaya*** **L.**

Family: Caricaceae

1. Introduction

Papaya, a native of tropical America, has now spread all over the tropical world. Papaya is grown mostly for fresh consumption or for production of proteolytic enzyme papain from the fruit latex. The ripe fresh fruits of papaya are sweet and delicious. The unripe fruits are also commonly used as vegetables for cooking. Papaya is a very wholesome fruit, and ranked second only to mango as a source of carotene. Papaya is liked for its attractive pulp color, flavor, succulence, and characteristic aroma.

It contributes greatly to the vitamin C component of the diet. Papaya cultivation for production of papain enzyme is a popular small scale industry. Brazil, Mexico, Thailand, Indonesia, and India are the leading producers of papaya. In addition, Hawaii, Taiwan, Puerto Rico, Peru, Bangladesh, and Australia also cultivate this crop commercially. Hawaii and Latin American countries also export papayas to the United States.

2. Composition (Per 100g of Edible Portion)

Constituent	*Composition*
Moisture (%)	89.6
Proteins (g)	0.5
Fat(g)	0.1
Carbohydrates (g)	9.5
Calcium (mg)	10.0
Phosphorus (mg)	10.0
Vitamin A (IU)	2020
Vitamin C (mg)	40
Nicotinic acid (mg)	0.2
Riboflavin (mg)	0.25

Chemical Composition of Some Papaya Varieties Grown in India

Variety	*Vitamin A (IU)*	*Vitamin C (g%)*	*Sucrose (%)*	*Glucose (%)*	*Fructose (%)*	*Crude Fiber (%)*	*Phosphorus (mg %)*	*Potassium (mg*	*Calcium (mg %)*
Coorg Honey Dew	3649	66.6	1.49	5.23	3.18	0.44	4.10	0.46	14.70
Pink Flesh (selfed)	4715	88.2	1.41	3.13	4.16	0.52	5.06	0.60	14.40
Pink Flesh Sweet	3399	63.2	2.47	4.09	3.06	0.32	5.01	0.40	12.93
Solo Large Sweet (selfed)	3749	93.8	1.10	4.36	3.00	0.38	4.06	0.40	8.03
Solo (open pollinated)	6347	117.1	1.12	4.15	3.02	0.36	4.28	0.49	14.74
Solo Small (open pollinated)	5381	125.9	0.95	3.34	3.74	0.37	4.75	0.48	14.72
Solo Small (open pollinated)	4965	91.7	0.99	2.91	3.26	0.34	4.40	0.47	21.04
Solo Yellow Sweet (selfed)	5110	82.2	0.52	3.06	4.19	0.37	4.35	0.40	14.47

Contd...

Contd...

Variety	Vitamin A (IU)	Vitamin C (g%)	Sucrose (%)	Glucose (%)	Fructose (%)	Crude Fiber (%)	Phosphorus (mg %)	Potassium (mg	Calcium (mg %)
Sunrise (open pollinated)	1600	46.3	0.48	4.16	3.71	0.49	4.45	0.36	11.40
Sunrise (selfed)	1599	71.3	1.96	3.58	2.34	0.39	4.96	0.43	17.40
Thailand	2199	46.6	1.82	4.44	2.64	0.57	4.39	0.33	8.41
Washington	5115	78.1	1.81	5.24	2.90	0.34	7.04	0.63	12.03

3. Suitable Cultivars

In India, Washington, Honey Dew, Coorg Honey Dew, CO1, CO2, CO4, Pusa Delicious, Pusa Majesty, Pusa Giant, Pusa Nanha, and Pusa Dwarf are grown extensively. Coorg Honey Dew is mostly hermaphrodite. CO2, though dioecious, is a good table type with high papain yield. CO2 is recommended exclusively for papain production, while CO1 is a duel-purpose type. Pusa Majesty is a good keeper, with little spoilage in transport. Pusa Giant withstands strong winds and is considered suitable for canning.

4. Maturity Indices

Change of skin color from dark-green to light green with some yellow at the blossom end (color break). Papayas are usually harvested at color break to ¼ yellow for export or at ½ to ¾ yellow for local markets. Flesh color changes from green to yellow or red (depending on cultivar) as the papayas ripen. A minimum TSS of 11.5 per cent is required by the market standards.

Maturity Stages

5. Quality Indices

Papayas picked ¼ to full yellow taste better than those picked mature-green to ¼ yellow because they do not increase in sweetness after harvest.

Uniformity of size and color; firmness; free from defects such as sunburn, skin abrasions, pitting, insect injury, and blotchy coloration and no decay.

6. Postharvest Management

Harvesting

The fruit is cut off with a thin blade knife and placed stem end down in a box padded with paper or wood wool to prevent bruising. The fruits are packed in bamboo baskets for transport over short distances and in wooden crates for long distances. Up to six fruits are packed in to each crate by providing proper cushions between the individual fruits. Papaya which is yellow but firm is best suited for transport.

Storage

Before storage to clean the fruit from dirt, soil, insect excreta and sap fruit washing machine developed at PAU, Ludhiana is found useful.

(*a*) Optimum Temperature

13°C for mature-green to ¼ yellow papayas

10°C for partially-ripe (¼ to ½ yellow) papayas

7°C for ripe (>½ yellow) papayas

(*b*) Optimum Relative Humidity

90-95 per cent

(*c*) Rates of Respiration/CO_2 Production

Temperature	7°C	10°C	13°C	15°C	20°C
ml CO_2/kg hr	3–5	4–6	7–9	10–12	15–35

To calculate heat production multiply ml CO_2/kg·hr by 122 to get kcal/metric ton/day.

(*d*) Rates of Ethylene Production

Temperature	7°C	10°C	13°C	15°C	20°C
ml C_2H_4/kg hr	0.1-2	0.2-4	0.3-6	0.5-8	1-15

(*e*) Responses to Ethylene Production

Exposure to 100 ppm ethylene at 20 to 25°C and 90-95 per cent relative humidity for 24-48 hours results in faster and more uniform ripening (skin yellowing and flesh softening, but little or no improvement in flavor) of papayas picked at color break to ¼ yellow stage.

(*f*) Responses to Controlled Atmospheres (CA)

Optimum CA 3-5 per cent O_2 and 5-8 per cent CO_2

Benefits of CA include delayed ripening and firmness retention.

Postharvest life potential at 13°C: 2-4 weeks in air and 3-5 weeks in CA, depending on cultivar and ripeness stage at harvest.

Exposure to O_2> levels below 2 per cent and/or CO_2 levels above 8 per cent should be avoided because of the potential for development of off-flavors and uneven ripening.

However for short term cool storage evaporative cooled storage structure developed at CIPHET, Ludhiana can be effectively utilized.

7. Postharvest Diseases and Disorders

Physiological and Physical Disorders

Skin Abrasions

Result in blotchy coloration such as green islands (areas of skin that remain green and sunken when the fruit is fully-ripe) and accelerate water loss. Abrasion and puncture injuries are more important than impact injury for papayas.

Cold Storage/Low Temperature Injury

Symptoms include pitting, blotchy coloration, uneven ripening, skin scald, hard core (hard areas in the flesh around the vascular bundles), water soaking of tissues, and increased susceptibility to decay. Increased alternaria rot was observed in mature-green papayas kept for 4 days at 2°C, 6 days at 5°C, 10 days at 7.5°C, or 14 days at 10°C. Susceptibility to chilling injury varies among cultivars and is greater in mature- green than ripe papayas (10 vs. 17 days at 2°C; 20 vs. 26 days at 7.5°C).

Heat Injury

Exposure of papayas to temperatures above 30°C for longer than 10 days or to temperature-time combinations beyond those needed for decay and/or insect control result in heat injury (uneven ripening, blotchy ripening, poor color, abnormal softening, surface pitting, accelerated decay). Quick cooling to 13°C after heat treatments minimizes heat injury.

Heat Treatments for Insect Control

Hot Water Treatment

30 minutes at 42°C followed within 3 minutes by a 49°C dip for 20 minutes.

Vapor Heat Treatment

Fruit temperature is raised by saturated water vapor at 44.4°C until the center of the fruit reaches that temperature, and then held for 8.5 hours.

Forced hot air treatment: 2 hours at 43°C + 2 hours at 45°C + 2 hours at 46.5°C + 2 hours at 49°C.

Pathological Disorders

Anthracnose

Anthracnose caused by *Colletotrichum gloesporioides*, is a major cause of postharvest losses. Latent infections of unripe papayas develop as the fruits ripen. Lesions appear as small, brown, superficial, watersoaked lesions that may enlarge to 2.5 cm or more in diameter.

Black Stem-end Rot

Black stem-end rot caused by *Phoma caricae-papayae* attacks fruit pedicel. After harvest, the disease lesion on fruits appear in the stem area which becomes dark-brown to black. Another stem-end rot is caused by *Lasiodiplodia theobromae*.

Phomopsis Rot

Phomopsis rot caused by *Phomopsis caricae papayae* begins in the stem end or a fruit skin wound and can develop rapidly in ripe fruits; invaded tissue softens and darkens slightly.

Phytophthora Stem-end Rot

Phytophthora stem-end rot caused by *Phytophthora nicotianae* var. parasitica begins as water-soaked areas followed by white mycelium that become encrusted.

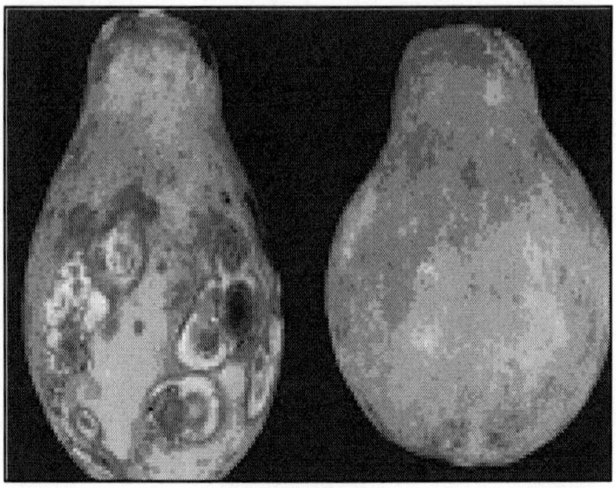

Alternaria Rot

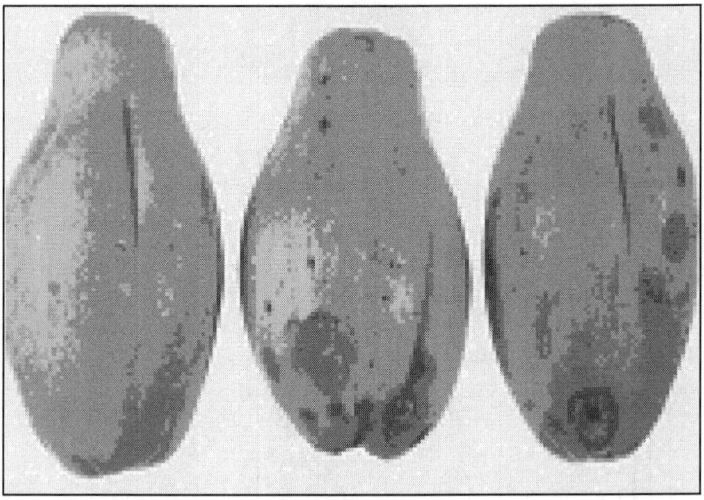

Anthracnose Rot

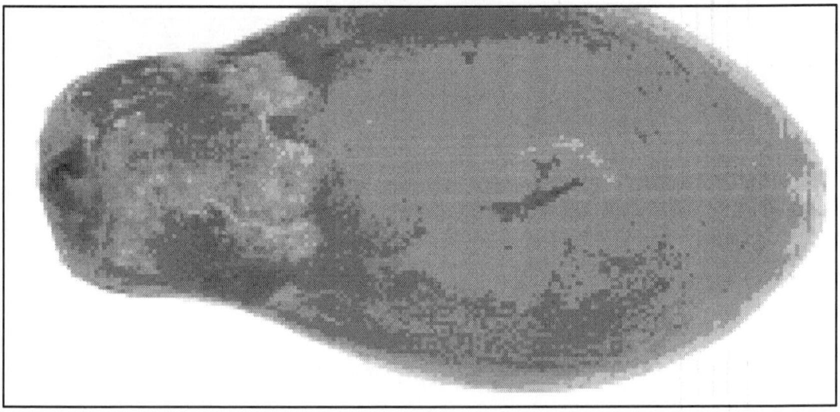

Phytopthora Stem End Rot

Alternaria Rot

Alternaria rot caused by *Alternaria alternata* follows chilling injury of papayas kept at temperatures below -12°C.

Control Strategies

- ☆ Careful handling to minimize mechanical injuries
- ☆ Prompt cooling and maintenance of optimum temperature and relative humidity throughout postharvest handling operations.
- ☆ Application of fungicides, such as thiabendazole (TBZ).
- ☆ Dipping in hot water at 49°C for 20 minutes.

5

Apple

Botanical Name: *Malus domestica* **Borkh**

Family: Rosaceae

1. Introduction

Apple is a highly remunerative deciduous fruit grown in temperate regions. It is believed to have originated from the hybridization between *Malus sylvestris* and other *Malus* species, with its original home said to be a region south of the Caucasus. The major apple-producing countries in

order of production are the former USSR, China, the United States, Germany, France, Italy, Turkey, Iran, Argentina, Japan, India, Hungary, Poland, the Korean Republic, Chile, Brazil, Spain, and Yugoslavia.

2. Chemical Composition (Per 100g Edible Portion)

Constituent	Composition
Water (per cent)	84.6
Food energy (kJ)	59
Protein (g)	0.2
Fat (g)	0.5
Carbohydrate total (g)	13.4
Fiber (g)	1.0
Calcium (mg)	10
Phosphorus (mg)	14
Iron (mg)	0.66
Chlorine (mg/100g)	
Ascorbic acid (mg)	1.0

3. Suitable Cultivars

There are about 5000 cultivars of apple grown all over the world, of which only a few could reach the status of commercial cultivars. The new cultivars are generally more resistant to diseases and more productive than established cultivars.

In India following cultivars for different apple-growing states of India: Starking Delicious, Granny Smith, Yellow Newton, Rich-a-Red, Red Gold, McIntosh, Red June, King of Pippins, Golden Delicious, Tydeman's Early for Himachal Pradesh, Ambri, Lal Ambri, Maharaji, Red Delicious, Sunehari, Golden Delicious, Benonic, Irish Peach, Cox's Orange Pippin, Kerry Pippin, Lal Cider, Apirough for Jammu and Kashmir, and Rymer, Buckingham, Fanny, Cortland, and Early Shanberry for Uttar Pradesh. Of late, "Delicious" varieties have gradually replaced other varieties of apple and contribute an increasing proportion of world production.

4. Maturity Indices

Indices of harvest maturity of apples are based largely on color (external and internal), flesh firmness, composition (starch, sugar, and acid), mechanical properties (rupture force, modules of elasticity), ease of separation from spurs, and days from full bloom to harvest.

Color change from dark green to light green or yellowish-green. Firmness of 14 force, 20 to 40 per cent of cortex clear of starch. Generally about 135 to 150 days from full bloom.

Non-destructive Methods for Determining Maturity of Fruits Developed at CIPHET, Ludhiana

Eating quality and postharvest shelf life of ripe fruits depends on its maturity at harvest. Change of peel color and total soluble solids (TSS) are indicative parameters to measure maturity of most of the fruit. A maturity index is defined based on TSS and colour values. The model was developed to predict maturity using colourimeter. Colour and maturity index chart were also developed to determine maturity. This technique can be employed to sort the fruits at well-organized markets and in processing plant.

5. Quality Indices

- Firmness, crispness, lack of mealiness
- Flavor, including soluble solids, titratable acidity and flavor volatiles.
- Free from defects such as bruising, decay, stem or blossom-end cracks, bitter pit, scald, internal browning, or shrivel.

6. Postharvest Management

Harvesting Methods

The ultimate use of the apple fruits decides the method to be used in harvesting. The most commonly used harvest method for fresh-market and processing apples is by hand.

Grading

Apples are graded for size and quality under groups A, B and C, depending on their color, regular shape, freedom from injuries, blemishes, diseased spots, etc.. A and B grades are sent to market; C-grade fruits are not generally marketed through the fresh fruit trade. Grades A and B are further subdivided by size, the grades depending on the equatorial diameter of the fruit. Fruits meant for export are further designated as extra fancy,

fancy class I, and fancy class; II. Mechanical graders are also used to provide uniform standards of size grades.

Before packing fruits can be graded according to size by using fruit grader developed by PDKV, Akola.

Packaging

Packaging is one of the major factors which influences the quality of fruit when it reaches the consumer.

Corrugated fiber board (CFB) cartons are replacing wooden boxes for packaging of apple. CFB cartons are capable of withstanding various transportation hazards both on muleback and by trucks. They cause minimum bruise damage (3.2-3.4 per cent) besides reducing loss in fruit weight. They also reduce pressure on forests. Plastic crates have also recently been recommended and introduced in the apple trade in Himachal Pradesh, India, to a limited extent, for field boxes for collection of fruit from orchards, stacking in cold storage, carriage to nearby markets, and to supply fruit to processing plants.

Storage

The storage life of apples depends on cultivar, production area, cultural practices, climatic conditions, maturity, handling, and transportation. For maximum storage, apples must be harvested when matured but not fully ripe. Immature apples have poor eating quality and are likely to shrivel in storage. Apples picked too mature will develop breakdown prematurely and have short storage life.

(*a*) Optimum Temperature

0°C ± 1°; Freezing temperature: -1.7°C

(*b*) Optimum Relative Humidity

90-95 per cent RH

(*c*) Rates of Respiration/CO_2 Production

Temperature	0°C	5°C	10°C	20°C
ml CO_2/kg hr	3 6	4–8	7–12	15–30

To calculate heat production multiply ml CO_2/kg·hr by 122 to get kcal/metric ton/day.

(*d*) Rates of Ethylene Production

Temperature	0°C	5°C	10°C	20°C
µl/kg hr	1-10	2-25	5-60	20-150

(*f*) Responses to Ethylene

Ethylene stimulates ripening. Mixed results on the benefits of scrubbing ethylene from storage rooms, depending on harvest maturity, and duration and type of storage (air or CA).

Responses to Controlled Atmospheres (CA)

Fruit to be stored longer than one month benefit from CA storage in terms of retention of flesh firmness, acidity, and skin color. CA storage potential is up to 10 months (vs. 6 months in air).

Recommended atmospheres: 1 to 3 per cent O_2 + 1.5 to 3 per cent CO_2

7. Postharvest Disorders, Diseases and their Control

Physiological Disorders

Shrivel

Golden Delicious apples are particularly susceptible to water loss. Weight loss can be as high as 3 to 6 per cent. Rapid cooling, storage of fruit with plastic bin liners, and well-designed refrigeration equipment reduces water loss.

Bruising

Can be excessive, especially for Golden Delicious where bruises are more visible. Gentle handling is important.

Bitter Pit

Sunken brown spots on the skin, especially at the calyx end, related to low calcium concentrations in the apple. Best controlled by calcium sprays prior to harvest and calcium dips prior to cold storage. Apply field sprays under rapid drying conditions to avoid russeting. Reduced incidence with controlled atmosphere storage is found.

Superficial Scald

It is the browning of the skin which develops in cold storage. Susceptibility of Golden Delicious to this disorder is low. Controlled atmosphere storage delays onset.

Controlled Atmosphere Damage

Oxygen levels below 1 per cent and CO_2 above 15 per cent induces off-flavors due to fermentative metabolism. Other symptoms of CO_2 injury include partially sunken brown lesions on skin or internal browning and cavities.

Pathological Disorders

Moldy Core

Caused by several fungi including *Alternaria* sp., *Fusarium* sp., *Aspergillus* and *Penicillium*. Golden Delicious apples are particularly susceptible because of the open or deep sinus cavity. Drenching can increase the incidence of moldy core.

Blue Mould and Grey Mould

The two most important postharvest diseases of Golden Delicious

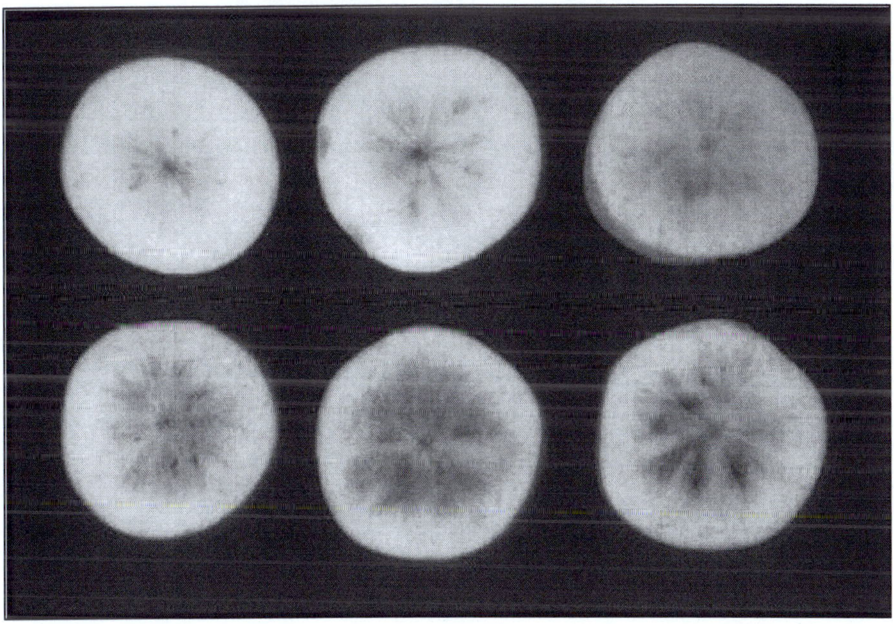

Low O_2 Injury

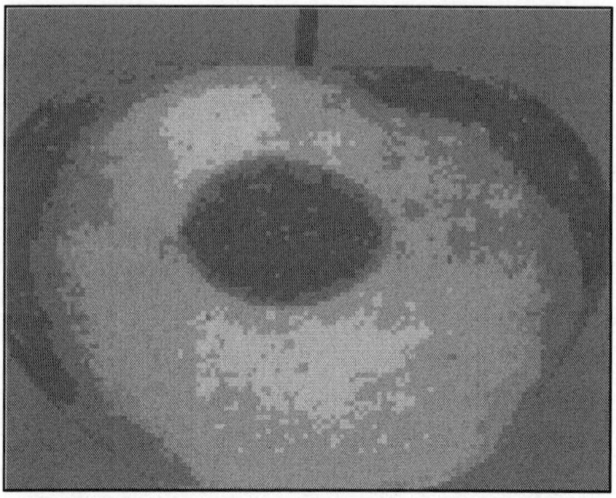

Blue Mould

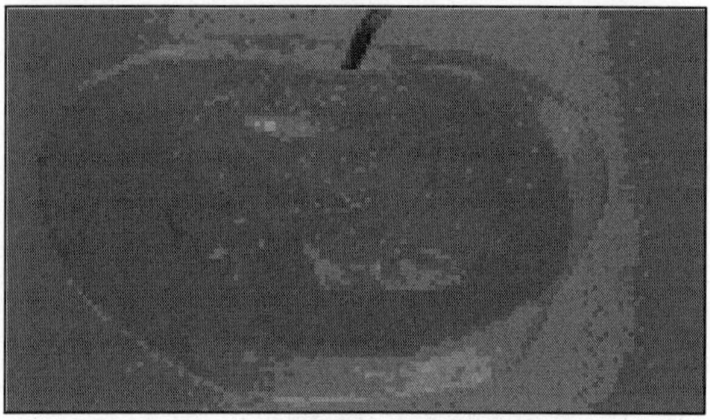

Grey Mould

apples are caused by *Penicillium expansum* and *Botrytis cinerea.* Both fungi are wound pathogens. Sanitation is critical to control of these diseases. Drenching can spread spores of *Penicillium* and *Botrytis* to wounds from harvest operations. Use of fungicides during drenching may reduce decay.

6

Pear

Botanical Name: ***Pyrus communis*** **L**

Family: Rosaceae

1. Introduction

Pears are grown in temperate areas. These are deciduous, cold-requiring fruit crops which can also grow in tropical lowlands but never set fruit there, since they need a certain number of hours with temperatures below 7°C to replace the dormant period of the temperate climates. In India pear is grown in J&K, Himachal Pradesh, Uttranchal, and sub tropical regions of Punjab, Haryana, Uttar Pradesh. Pears are consumed primarily as fresh fruit. However pear is also very popular for processing in to RTS, preserves, jams, jelly.

2. Composition (Per 100g of Edible Portion)

Constituent	Composition	
	Raw	Dried
Water (per cent)	87.7	26.0
Food energy (kJ)	264.60	1369.80
Protein (g)	0.7	5.4
Fat (g)	0.4	3.2
Carbohydrate total (g)	15.8	64.1
Fiber (g)	1.4	–
Calcium (mg)	13.0	61.0
Phosphorus (mg)	16.0	84.0
Iron (mg)	0.3	2.3
Vitamin A (iu)	20.0	120.0
Thiamin (mg)	0.02	0.2
Riboflavin (mg)	0.04	0.32
Nicotinic acid (mg)	0.1	1.1
Ascorbic acid (mg)	4.0	12.0

Extracts of different parts of the plant have shown variable antibacterial action. Fresh pear juice exhibited good activity against *Micrococcus pyrogenes* and *Escherichia coli*. An aqueous extract of the leaves was active against some strains of *Escherichia coli*. Chlorogenic acid is present in the vegetative parts of the tree. The leaves contain arbutin, isoquercitin, sorbitol, astragalin, and tannin.

3. Suitable Cultivars

Pear varieties belong to 3 groups–European, Asian and hybrids. The varieties recommended for different states are listed below:

Pear Varieties for High Hills

Early	Mid-season	Late
Early China,	Barlett,	Conference
Laxton's	Starking Delicious,	Comice,
Fertility (P),	Max-Red Barlet,	Winter Nellis, Clapp's
Seckel	Dr. Jule's Guyot	Winter Nellis, Clapp's Favourite Flemish Beauty

Himachal Pradesh

High hills: The pears are classified as early, mid-season and late ripening. They are: Mid, low hills and valley areas: Pears grown in these areas are : Patharnakh (Sand Pear), Keiffer and China Pear.

Jammu and Kashmir

Pear Varieties for Temperate Areas

Early	*Mid-season*	*Late*
China pear,	Citron-do-Carme, Clapp's	Hardy
Beurre-de-Amanlis	Favorite, Doyenne Bussoch Genta Drauard Fertility China Sand Pear, Willian Barlett	Viear of Winkfield

Uttar Pradesh

The varieties grown are:

High hills: Max-red Barlett, William Barlett, Confernce, Hardy, Winter Nelis, Clapp's Favourite, Flemish Beauty and Comice.

Lower hills and plains: Patharnakh, Buggugosha, Punjab Beauty and LeConte

Composition of Edible Portion of Important Types of Pears Grown in India

Constituent	*Bagugosha*	*Kashmiri Nakh*	*Country*	*Kieffer*
Moisture (g/100 g)	86.5	83.6	86.0	86.3
Protein (g/100 g)	0.4	0.2	0.2	0.2
Fat (g/100g)	0.1	0.3	0.1	0.2
Minerals (g/100 g)	0.3	0.4	0.3	0.2
Fiber (g/100 g)	2.1	0.6	1.0	1.4
Other carbohydrates (g/100 g)	10.6	14.9	12.4	11.7
Calcium (mg/100 g)	20.0	20.0	6.0	10.0
Phosphorus (mg/100 g)	20.0	20.0	10.0	10.0
Iron (mg/100 g)	1.5	1.0	1.0	0.4
Vitamin A (I.U./100 g)			14.0	9.0
Thiamin (mg/100 g)	–		0.02	0.03
Riboflavin (mg/100 g)	–	–	0.03	0.02
Nicotinic acid (mg/100 g)	0.2		0.2	
Vitamin C (mg/100 g)	1.0	3.0		7.0

4. Maturity Indices

Pears attain best eating quality if picked at the mature-green stage and ripened off the tree. They become mealy if ripened on the tree.

They can be determined by combined flesh firmness/soluble solids content (SSC) index that is further modified by fruit size and skin color.

Minimum SSC	*Maximum Flesh Firmness (Pounds-Force)**	
	Fruit Diameter 2 3/8"–2 ½"	*Fruit Diameter >2 ½"*
<10 per cent	19.0	20.0
10 per cent	20.0	21.0
11 per cent	20.5	21.5
12 per cent	21.0	22.0
13 per cent	[No maximum]	

* Penetration force with 8 mm (5/16 inch) tipM.

5. Quality Indices

Fruit shape, size, and free from insect damage, mechanical injuries (impact, compression, and/or vibration bruising), decay, and other defects.

Sweet taste, pleasant aroma, and juicy, buttery texture are desired eating characteristics of ripe pears (flesh firmness range between 2 and 4 force).

6. Postharvest Management

Harvesting

The most commonly used harvest method for fresh-market and processing pears is by hand.

Precooling

If the pears are going into long-term storage, they need to be cooled down or precooled to storage temperature within 48 h of harvest. Forced-air cooling is the most suitable method, as hydrocooling has a detrimental effect on the pear skin. Once precooled, the pears can be either stored or packed.

Ripening

Although the pears can ripen while still attached to the tree, a generally accepted commercial practice is to pick the fruit before the onset of respiratory climacteric rise. This practice aids in increasing shelf life of fruit during storage and transport. Pears may require special ripening conditions for best quality. They ripen best and with more aromatic flavors at 18.3-21.1°C than at either lower or higher temperatures. In fact, some cultivars will not ripen if the temperature is above 24°C (25). Once a fruit is ripe, its shelf life is very short but may be prolonged by refrigerated storage.

Irradiation

Pears are irradiated with-γ-rays at 500,1000, and 1500 Gy or x-rays at 40,60, and 100 Gy before cold storage in CA or normal atmosphere. Ripening is accelerated after irradiation. Pear fruits could be stored satisfactory for 7 months at 0-0.5°C, 92 per cent RH, in 2 per cent CO_2 and 2 per cent O_2.

Storage

Before storage pear fruit are washed and for this job fruit washing machine developed at PAU, Ludhiana can be effectively utilized.

In general pears can be best stored at –1.7°C to –0.6°C

(*a*) Optimum Temperature

-1 to 0°C

(*b*) Optimum RH

90 to 95 per cent Relative Humidity

(*c*) Rates of Respiration/CO_2 Production

Temperature	0°C	2°C	5°C	20°C
ml CO_2/kg hr:	2–3	4–5	6–8	15–35

To calculate heat production multiply ml CO_2/kg hr by 122 to get kcal/metric ton/day.

(*d*) Rates of Ethylene Production

Temperature	-1 to 0°C	5°C	10°C	20°C
µl C_2H_4/kg hr:	0.1–0.5	2–4	5–15	20–100

(*e*) Responses to Ethylene

Mature-green pears can be treated with ethylene at harvest (100 ppm ethylene) for 24 to 48 hours at 20 to 25°C to ensure uniform ripening within 4 to 6 days once fruit are warmed for ripening. Fruit, which has been cold stored for 3 weeks, will ripen uniformly without ethylene treatment. These pears can also be ripened in the presence of ethylene gas (100 ppm ethylene) at 18 to 22°C.

100 ppm C_2H_4

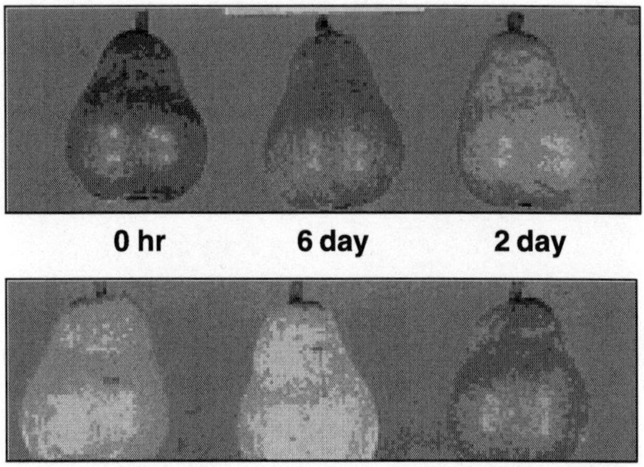

0 hr 6 day 2 day

1 day 12 hr Ambient 6 day

(*f*) Responses to Controlled Atmospheres(CA)

Optimum CA 1 to 3 per cent O_2 + 0 to 3 per cent CO_2; 1.5 to 2 per cent O_2 + 1 to 5 per cent CO_2 are recommended for long-term storage of early- and mid- season harvested fruits. For late-season pears, CO_2 should be kept below 1 per cent because of the fruit's greater sensitivity to CO_2 injury (core and flesh browning).

CA conditions slow the rates of respiration, ethylene production, color change from green to yellow, and softening of pears. Scald development and decay incidence are suppressed under CA. Storage potential of 'Bartlett' pears at -1 to 0°C and 90-95 per cent RH can be as long as 3 months in air and 6 months in CA.

7. Postharvest Disorders and Diseases

Physiological Disorders

Senescent Scald

Dark-brown skin discoloration begins as small spots and develops into large areas of skin turning brown after long term storage; brown skin may slough off. Prompt cooling and maintenance of optimum fruit temperature minimize this disorder. This disorder occurs when fruit become senescent.

Superficial Scald

Diffuse brown skin discoloration occurs after several months of storage and during ripening after storage. Scald inhibitors, such as ethoxyquin can be used on pears that are stored longer than 3 months. Controlled-atmosphere storage delays scald incidence and severity.

Watery Breakdown

Losses from this disorder result from a rapid enzymatic flesh breakdown (excessive softening) that occurs sometimes during storage, but mostly during subsequent ripening. Prompt and thorough cooling and maintenance of optimum temperatures of -1 to 0°C minimizes losses due to watery breakdown.

Pink Calyx

Cool growing temperatures can result in a "pink calyx" fruit breakdown and premature ripening of pears.

CA-related

Exposure of pears to unfavorable atmospheric compositions (levels of O_2 below and/or CO_2 above those indicated above as optimum CA) can induce physiological disorders and failure to ripen upon removal from CA conditions. CA-stressed pears exhibit flesh browning, develop cavities in damaged tissues, and undergo ethanolic fermentation, which results in accumulation of acetalydehyde, ethanol, and ethyl acetate, and development of off-flavors.

Physical Disorders

Vibration Bruising

Symptoms include brown spots or bands on the skin where rolling, rubbing, or vibration occurred, and damage is usually not visible below the surface. This problem can be minimized by immobilizing the pears (such as tight- fill packing) or by packaging them into plastic bags during transport.

Impact Bruising

Brown discoloration of the flesh results from impact bruising; severity increases with the height of the fruit drop. Partially-ripe and ripe pears are more susceptible to impact bruising than mature-green pears.

Pathological Disorders

Blue Mould Rot

Blue mould rot is caused by *Penicillium expansum* which is a wound parasite that readily colonizes cuts and punctures on pears. Avoiding mechanical injuries is the best way to avoid this fungus.

Gray Mould Rot

Gray mould rot is caused by *Botrytis cinerea*. Infections occur during blossoming and remain latent until the pears begin to ripen at which time

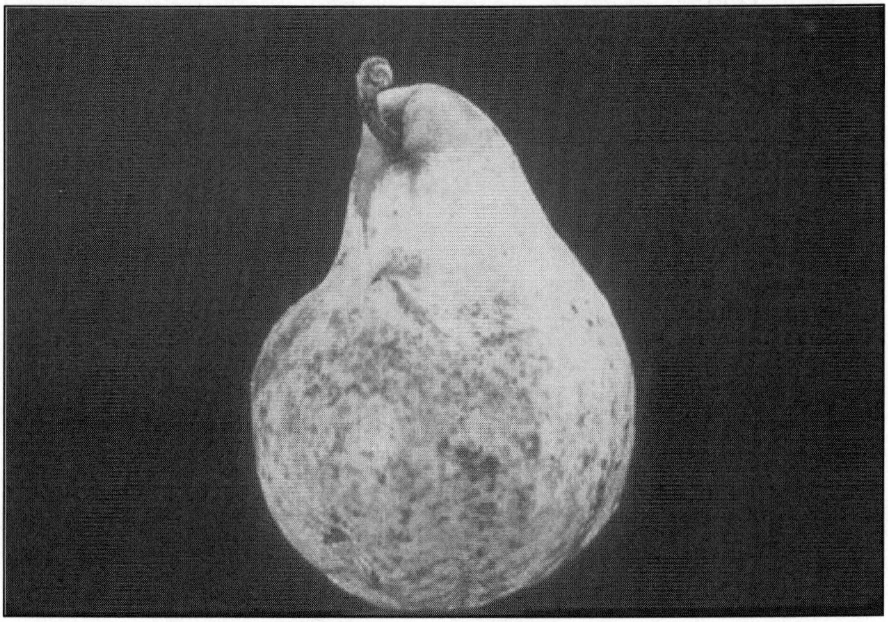

Abrasion Damage

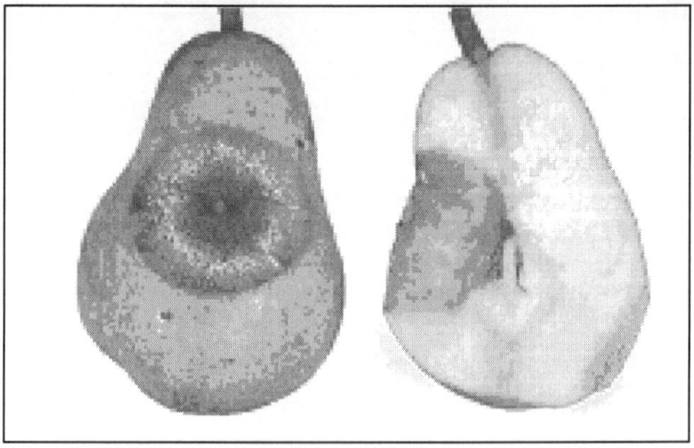

Blue Mould Rot

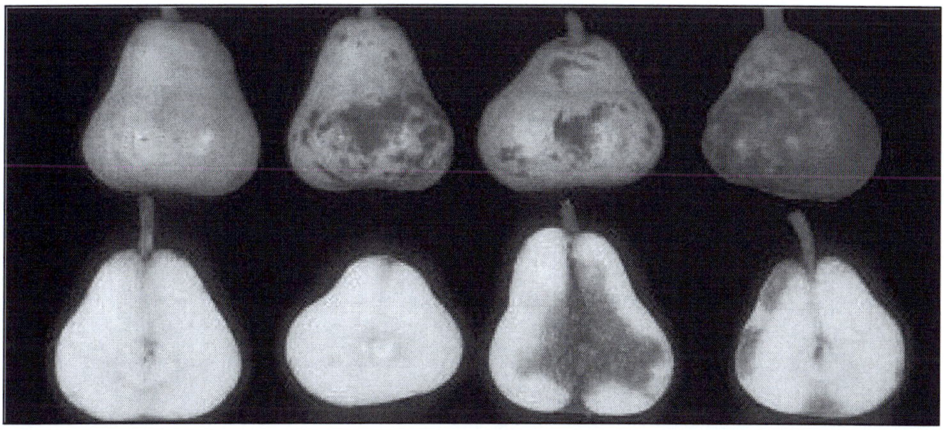

Air | 0.25% O_2 | 80% CO_2 | 0.25% + O_2 80% CO_2

After 3 Days at 20°C

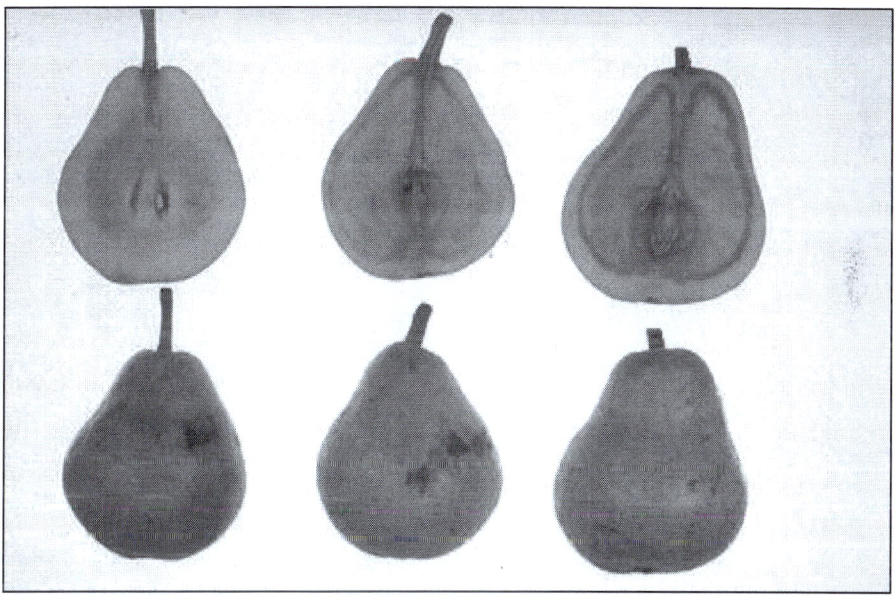

Freezing Damage of Pear

the fungus can grow into the fruit, especially at the calyx and stem ends. Botrytis can also enter the fruit through wounds created during postharvest handling. Control strategies include minimizing mechanical damage, prompt cooling and maintaining optimum temperature during transport and storage.

7

Plum

Botanical Name: ***Prunus domestica***

Family: Rosaceae

1. Introduction

Plum is one of the most important stone fruit of the world and appreciated for its excellent quality. Plums with high sugar content and firm flesh are dried without the removal of stone and are called prunes. The fruit is a rich source of sugars. The colour of the fruit is mostly contributed by anthocyanins, which are located in the epidermal layer. Carotenoid also contributes to the colour of some plum. The fruit is a good source of minerals such as K, Na, Ca, Mg, Fe and Zn. The plum is considered to be a climacteric fruit. The respiration rate is high during growth. As maturity approaches, it decreases to a preclimacteric minimum and increases irreversibly to a maximum during ripening. At this stage, the fruit is soft and sweet, with a characteristic flavor and is ideal for eating. Subsequently, senescence sets in, whereupon the respiration rate decreases and the fruit becomes overripe.

2. Composition (Per 100 g edible portion)

The nutrient and mineral composition of plums varies according to variety and the environment under which the plums are grown. General composition includes

Constituent	*Composition*	*Constituent*	*Composition*
Moisture (%)	85.2	Vit. C(mg)	10.0
Energy value (kJ)	230	Vitamin E(mg)	0.65
Carbohydrates (g)	11.0	Sodium(mg)	0
Dietary fiber (g)	2.0	Potassium(mg)	172
Protein (g)	0.8	Calcium(mg)	4.0
Total fat (g)	0.6	Magnesium(mg)	7.0
Nicotinic acid (mg)	0.5	Phosphorus(mg)	10.0
Riboflavin (mg)	0.10	Iron(mg)	0.1
Thiamine (mg)	0.04	Copper(mg)	0.04
Folic acid (mg)	2.0	Zinc(mg)	0.10
Vitamin A(mg)	192.0		
Vitamin B(mg)	0.08		

3. Suitable Cultivars

European Plum and prunes: Queenston, California Blue, Washington, Early Italian, Grand Duke

Japanese cultivar and hybrids: Early Golden, Beauty, Satsuma

Low chilling cultivars: Satluj Purple, Kala Amritsari,

4. Maturity Indices

In most of the cultivars harvest date is determined by skin color changes that are described for each cultivar. A color chip guide can be designed to determine maturity for each cultivar.

A three tier maturity system is generally used.

1. Less Mature (Minimum Maturity)

2. Well-Mature
3. Tree Ripe

Measurement of fruit firmness is recommended for cultivars where skin ground color is masked by full red or dark color development before maturation.

Maximum maturity: Flesh firmness, measured with a penetrometer with an 8 mm diameter can be used to determine a maximum maturity index, which is the stage at which fruit can be harvested without suffering bruising damage during postharvest handling.

- ☆ Plums are less susceptible to bruising than most peach and nectarine cultivars at comparable firmness.

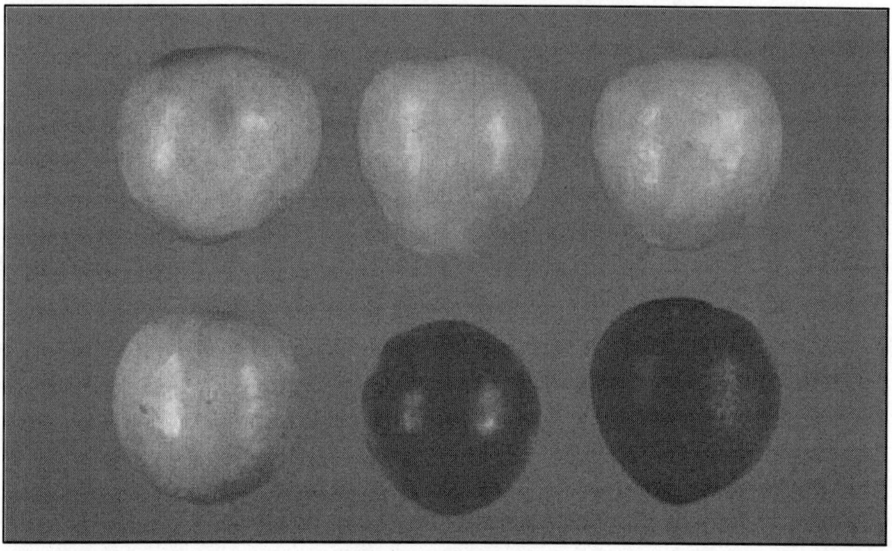

Ripeness Stages

5. Quality Indices

High consumer acceptance is for high soluble solids content (SSC) fruits. Fruit acidity, SSC/acidity ratio, and phenolic content are also important factors in consumer acceptance. Plums with 2-3 kgs force flesh firmness are considered "ready to eat".

6. Postharvest Management

Harvesting

Plums usually ripen unevenly on the tree. Therefore fruit should be harvested in two or three pickings. They should be picked by hand into buckets or baskets with padded liners or into picking bags. So far as possible, the peduncle should be allowed to remain attached to the fruit.

Packaging

Packages such as waxed fibre board cartons, parchment wraps and other specially treated packing materials retard moisture loss significantly.

Storage

Before storage plum fruits are washed for cleaning dirt, soil and insect excreta. Fruit washing machines developed at AICRP (PHT) centres is found very useful and efficient.

(*a*) Optimum Tempertures

-1.0 to 0°C

Freezing point varies depending on SSC.

(*b*) Optimum Relative Humidity

90-95 per cent R.H; an air velocity of approximately 2 Cum/min is suggested.

(*c*) Rates of Respiration/CO_2 Production

Temperature	0°C	10°C	20°C
ml CO_2/kg hr	1–1.5	4.2	8.2

To calculate heat production multiply ml CO_2/kg hr by 122 to get kcal/metric ton/day.

(*d*) Rates of Ethylene

Temperature	0°C	5°C	10°C	20°C
µl/kg hr	< 0.01–5*	0.02–15	0.04–60	0.1–200

The lower end of this range is for mature but unripe fruit; higher values are for ripe fruit.

(*e*) Responses to Ethylene

Most of the plums harvested at Well-Mature stage ripen properly without external ethylene application. Ethylene application to fruit harvested at the minimum maturity will only ripen the fruit more uniformly without speeding up the rate of ripening. However, for the slow ripening cultivars, external application of ethylene (100 ppm for 1-3 days at 20°C) is needed for even ripening.

(f) Responses to Controlled Atmospheres (CA)

The major benefits of CA during storage/shipment are retention of fruit firmness and ground color. Decay incidence has not been reduced by CA of 1-2 per cent O_2 + 3-5 per cent CO_2. CA conditions of 6 per cent O_2+ 17 per cent CO_2 are suggested for reduction of internal breakdown during shipment, but its effectiveness depends on cultivar, preharvest factors, market life and shipping time.

CA Effects

Air **1 per cent O_2 + 5 per cent CO_2**

7. Postharvest Disorders, Diseases and their Control

Physiological Disorders

Internal Breakdown or Chilling Injury

Internal Breakdown or Chilling Injury is characterized by flesh translucency, flesh internal browning, flesh mealiness, flesh bleeding, failure to ripen and flavor loss. These symptoms develop in plum and fresh prunes during ripening after a cold storage period. Fruit stored within the "killing temperature range" 2-8°C are more susceptible to this problem.

Pathological Disorders

Brown Rot

Brown rot is caused by *Monilia fructicola* is the most important postharvest disease of stone fruits. Infection begins during flowering and fruit rot may occur before harvest but often occurs postharvest. Orchard sanitation is essential to minimize infection sources. The preharvest fungicide application and prompt cooling after harvest are among the control strategies along with postharvest fungicide treatment.

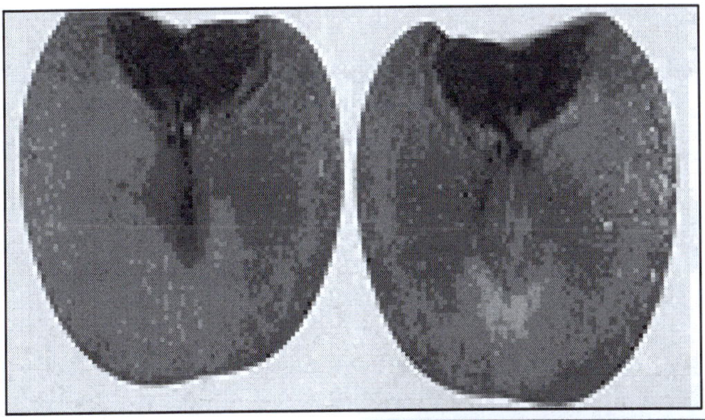

Alternaria alternata

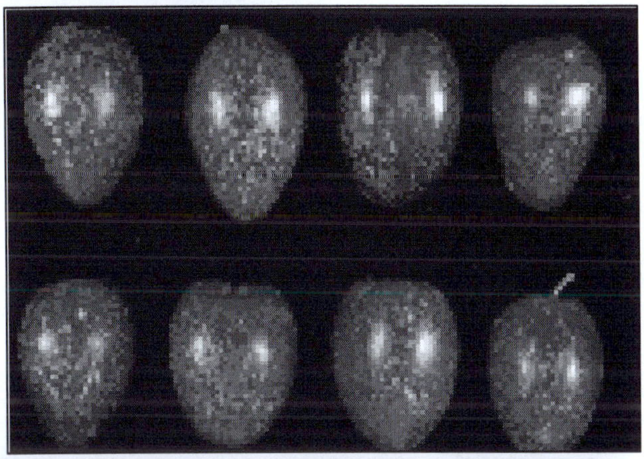

Sulfur Dioxide Injury

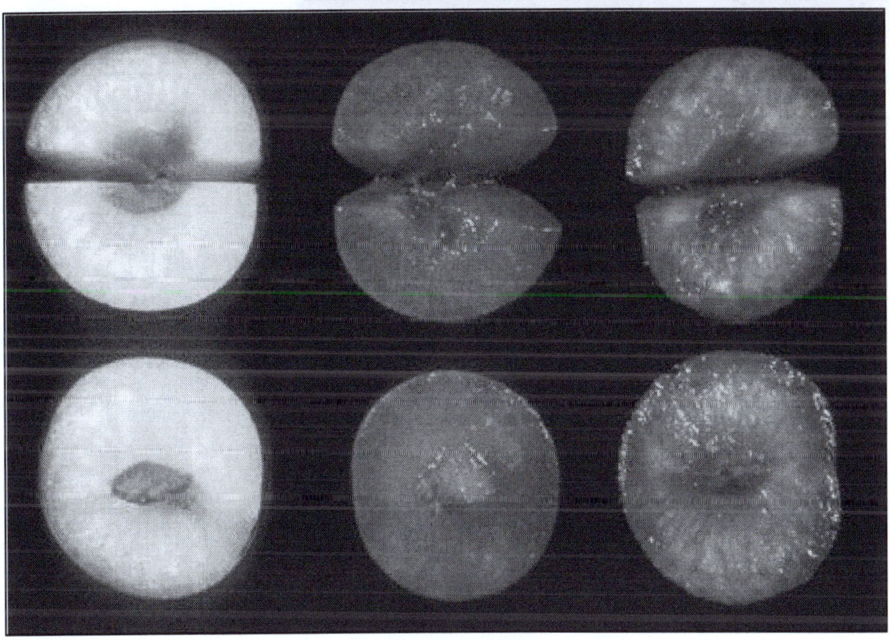

Chilling Injury

Gray Mould

Gray Mould is caused by *Botrytis cinerea* can be serious during wet spring weather. It can occur during storage if the fruit has been contaminated through harvest and handling wounds. Avoiding mechanical injuries and good temperature management are effective control measures.

Rhizopus Rot

Rhizopus Rot is caused by *Rhizopus stolonifer* and can occur in ripe or near ripe stone fruits kept at 20 to 25°C. Cooling the fruits and keeping them below 5°C is very effective against this fungus.

8

Peach and Nectarine

Botanical Name: *Prunus persica (L.)*

Family: Rosaceae

1. Introduction

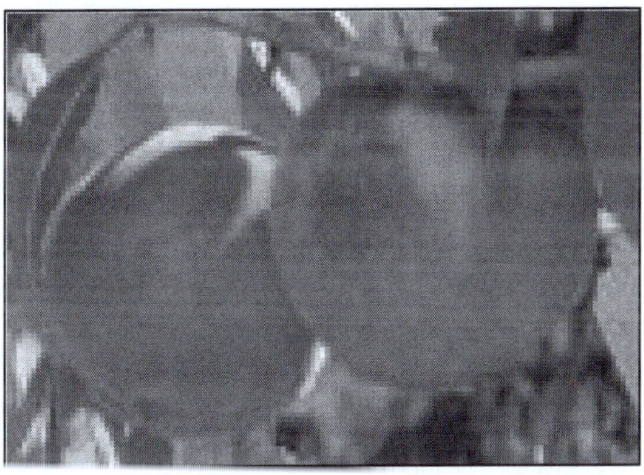

Peach, along with its smooth skinned mutant nectrine, is one of the most important temperate stone fruits grown in the world. In India peaches are grown in J&K, Himachal Pradesh, Uttarakhand, foothills of UP, Punjab, Haryana and Rajasthan. Peaches are highly valued as a table fruit for their attractive colour and palatability.

2. Composition (Per 100g of Edible Portion)

Constituent	*Composition*
Moisture (%)	86-89.1
Sugars (% Wt. Basis)	
Total	7.54-8.45
Reducing	2.45
Non reducing	6.35

Constituent	*Composition*
Proteins (%)	0.6-1.2
Titrable acidity (%)	0.63
Thiamine (μg)	0.02
Riboflavin (μg)	0.04
Niacin (μg/)	0.5
Vitamin C (mg)	1-27
Minerals (mg)	
Magnesium	21
Sodium	2.0
Potassium	453
Copper	0.06
Sulphur	26.0
Calcium	15
Phosphorus	41
Iron	2.4

3. Suitable Cultivars

Alexendra, Congress, Early Amber, Elberta, Flordasun, Maygold, Sharbati, Saharanpur, Prabhat, Florda prince, Shane-e-Punjab.

4. Maturity Indices

The harvest date is determined by skin ground color when it changes

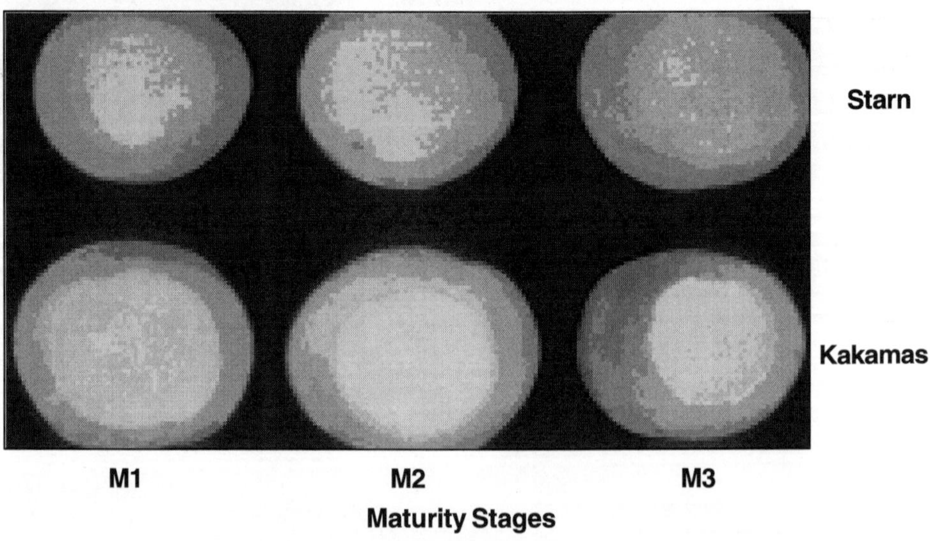

from green to yellow. A color chip guide can be used to determine maturity of each cultivar.

Measurement of fruit firmness is recommended in cultivars where skin ground color is masked by full red color development before maturation. Maximum maturity is considered as the flesh firmness at which fruits can be handled without bruising damage. It is measured with a penetrometer with an 8 mm tip. Bruising susceptibility varies among cultivars.

5. Quality Indices

High consumer acceptance is attributed to with high soluble solids content (SSC). The fruit acidity, SSC/acidity ratio, and phenolic content are also important factors in consumer acceptance. Fruit with 100-200 g force flesh firmness is considered "ready to eat". Fruit below 40-60 force are more acceptable to the consumer.

6. Postharvest Management

Harvesting and Packaging

All the fruits do not ripen at the same time on the tree. Earlier set fruits mature first and later set fruits mature later. Generally 3-4 pickings are done to complete the harvesting. During picking collect the fruits in baskets/plastic cartons after putting some dry grass or paper strips in it as a cushion to prevent injury or bruises to fruits. After harvesting transfer the fruits to some shady places with good aeration to make them cool down. Field heat of the fruits can be effectively removed by giving them 10-15 minute quick dip in cold water followed by surface drying the fruits in shed. This process slows down the ripening process of the fruit and help in extending the shelf life. Generally 2-4 kg capacity corrugated fibreboard boxes (CFB) are used for packaging of fruit.

Storage

(*a*) Optimum Temperature

-1 to 0°C

Freezing point varies depending on SSC from -3 to -2.5°C

(*b*) Optimum Relative Humidity

90-95 per cent R.H.; an air velocity of approximately 50 Cu m/min is suggested during storage.

(*c*) Rates of Respiration/CO_2 Production

Temperature	0°C	10°C	20°C
ml CO_2/kg hr	2–3	8–12	32–55

To calculate heat production multiply ml CO_2/kg hr by 440 to get Btu/ton/day or by 122 to get kcal/metric ton/day.

(*d*) Rates of Ethylene Production

< 0.01-5 µl/kg·hr (range)* at 0°C, 0.02-10 µl/kg·hr at 5°C, 0.05-50 µl/kg·hr at 10°C and 0.1-160 µl/kg·hr at 20°C

*The lower end of this range is for mature but unripe fruit; higher values are for ripe fruit.

(*e*) Responses to Ethylene

In general peaches and nectarines harvested at Well Mature (higher than minimum mature) will ripen properly without external ethylene application. Ethylene application to fruit harvested at the minimum maturity is useful only to ripen the fruit more uniformly without speeding up the rate of ripening.

(*f*) Responses to Controlled Atmospheres (CA)

The major benefits of CA during storage/transport are retention of fruit firmness and ground color. Decay incidence is not reduced by using CA 1-2 per cent O_2+ 3-5 per cent CO_2. CA conditions of 6 per cent O_2 + 17 per cent CO_2 are suggested for reduction of internal breakdown during transport but the efficacy is related to cultivar, preharvest factors, market life and transportation time period.

(g) Effects of Varieties and Cultural Practices on Postharvest Life

Maximum market life is obtained when fruit is stored at approximately 0°C. Maximum market life varies from 1-7 weeks for nectarine cultivars and from 1-5 weeks for peach cultivars. Because internal breakdown is the main limitation to market life, minimum postharvest life occurs when fruit is stored at 5°C. Cultural practices have an important role in determining fruit quality and storage potential. Leaf nitrogen content between 2.6-3.0 per cent is advised to obtain high red color development and maximum storage performance. Small size fruit grown in the outside canopy position have a longer market life than large size fruit grown in the inside position.

4. Disorders, Diseases and their Management

Physiological Disorders

Internal Breakdown or Chilling Injury

This physiological problem is characterized by flesh internal browning, flesh mealiness, flesh bleeding, failure to ripen and flavor loss. These symptoms develop during ripening after a cold storage period, thus, are usually detected by consumers. Fruit stored within the 2.2-7.6°C temperature range are more susceptible to this disorder.

Inking (Black Staining)

It is a comestic problem affecting only the skin of peaches and nectarines. It is characterized by black or brown spots or stripes. These symptoms appear generally 24-48 hours after harvest. Inking occurs as a result of abrasion damage in combination with heavy metals (iron, copper and aluminum) contamination. This occurs usually during the harvesting and handling operations, although it may occur in other steps during postharvest handling. Gentle fruit handling, short transport, avoiding any foliar nutrient sprays within 15 days before harvest, and following the suggested preharvest fungicide spray interval guidelines are some of the ways to reduce inking.

Pathological Disorders

Brown Rot

It is caused by *Monilia fructicola* and is one of the most important postharvest disease of stone fruits. Infection begins during flowering and fruit rot may occur before harvest but often occurs postharvest. Orchard sanitation to minimize infection sources, preharvest fungicide application, and prompt cooling after harvest are among the control strategies. Also, postharvest fungicide treatment may be used.

Gray Mould

It is caused by *Botrytis cinerea* can be serious during wet weather. It can occur during storage if the fruit has been contaminated through harvest and handling wounds. Avoiding mechanical injuries and good temperature management are effective control measures.

Wounding during Harvest and Handling

Brown Rot

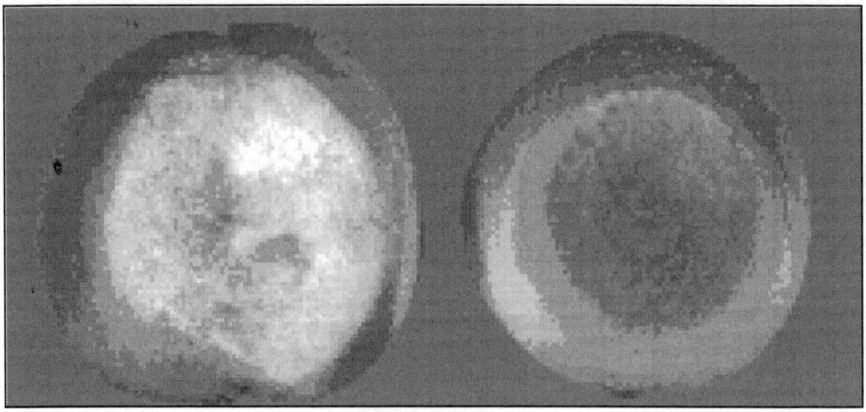

Grey Mould

Rhizopus Rot

It is caused by *Rhizopus stolonifer* and can occur in ripe or near ripe stone fruits kept at 20 to 25°C. Cooling the fruits and keeping them below 5°C is very effective against this fungus.

9

Pomegranate

Botanical Name: *Punica granatum* L

Family: Punicaceae

1. Introduction

Pomegranate is one of the favorite exotic fruits. Pomegranate is native to a region ranging from Iran to the Himalaya. Pomegranate is grown extensively in India, Afghanistan, Pakistan and Iran. India is world's largest

producer of pomegranate. In India it is grown in almost all parts but commercial cultivation exists only in Maharashtra, Gujarat, Karnataka, Rajasthan, U.P, Andhra Pradesh and Tamil Nadu. Pomegranate fruits vary in colour from pale yellow to red depending on varieties. The juicy pulp in the arils (seeds) varies from almost colourless to blood red. Pomegranate is a good source of carbohydrates and minerals such as Ca, Fe, and S and a moderate source of pectin. In India, the pomegranate has long been considered as a medicine of great importance. Now, scientific evidences are available which shows that fruits have restorative powers. According to new study, antioxidants present in pomegranate juice may help to reduce the formation of fatty acid deposits on the artery walls and also limits cell damage. The juice of fruit helps in combating genetic tendency towards hardening of the arteries (Atherogenesis) and related diseases, such as heart attack and strokes.

2. Composition (Per 100g of Edible Portion)

Constituent	*Composition*	*Constituent*	*Composition*
Moisture (%)	77.0–78.2	Thiamine (mg)	0.06
Proteins (g)	1.78–1.96	Riboflavin (mg)	0.1
Fat (g)	1.72–2.11	Niacin (mg)	0.3
Carbohydrates (g)	17.5–2	Vitamin C(mg)	16.0
Pectin (g)	0.47–0.55	Calcium (mg)	10.0
Total sugars (g)	6.2–9.0	Copper (mg)	0.17
Reducing sugars (g)	5.6–7.5	Phosphorus(mg)	70.00
Oxalic acid(mg)	14.0	Sodium (mg)	4.00
Energy Value (K cal)	65.0	Iron(mg)	0.30
		Potassium(mg)	17.1
		Magnesium(mg)	12.0
		Sulfur (mg)	25–28

3. Suitable Varieties

Ganesh, Bassein seedless, Jalore Seedless, GKVK-I, G-137. Wonderful, Granada, Ruby

Hybrids: Mridula, Ruby, Arkata, Bhagwa

4. Maturity Indices

- External red color (depending on cultivar)
- Red color of juice.
- Acidity of juice below 1.85 per cent

5. Quality Indices

- It should be free from growth cracks, cuts, bruises, and decay
- Skin color and smoothness
- Flavor depends on sugar/acid ratio which varies among cultivars. A soluble solids content above 17 per cent is desirable
- Tannin content below 0.25 per cent is desirable

Pomegranate fruits are susceptible to moisture loss and need to be stored at high humidity. After harvest the fruits are graded according to size, wrapped in paper and packed in corrugated boxes. In bulk storage, fruits are packed in layers each containing about 16-18 kg of fruit. Dry grass, rice straw or paper are used as cushioning material.

6. Postharvest Management

Harvesting and Packaging

It is advisable to harvest pomegranate fruits either early in the morning or in the evening hours. Fruits must be picked with clippers and kept in bags after harvest. But these fruits must not be pulled from branches and stem end must be cut close to the base. Keeping slightly longer stem end can cause damage to other fruits in handling and transport. During harvest care should be taken not to inflict any injury to the skin of the fruit. After harvest, fruits must be kept in a plastic crate without heaping too many layers. Care has to be taken that calyx or bracts should not be injured or broken during harvest or field storage. Plastic crates having fruits should be kept in shade and not in open sun. It is advisable to keep some cushioning material at the base of the crates.

Pomegranate being non-climacteric fruit should be picked when fully ripe. Harvesting of immature or over ripe fruits affects quality. Its fruits become ready for picking 140-150 days after fruit set. The calyx at the distal end of the fruit gets closed on maturity. Ripe fruits give a distinct sound of grains cracking inside when slightly pressed with thumb from out side.

Storage

Pomegranate fruits treated with calcium chloride 4 per cent and packed in CFB boxes (5 gauge).

(*a*) Optimum Temperature

At 5°C can be stored up to 2 months; longer storage should be at 7.2°C to avoid chilling injury.

(*b*) Optimum Relative Humidity

90-95 per cent; pomegranates are very susceptible to water loss resulting in shriveling of the skins. Storing fruit in plastic liners and waxing can reduce water loss, especially under conditions of lower relative humidity.

(*c*) Rates of Respiration/CO_2 Production

Temperature	5°C	10°C	20°C
ml CO_2/kg hr	2–4	4–8	8–18

To calculate heat production multiply ml CO_2/kg hr by 440 to get Btu/ton/day or by 122 to get kcal/metric ton/day.

(*d*) Rates of Ethylene Production

Less than 0.1 μl/kg·hr at 10°C or lower

Less than 0.2 μl/kg·hr at 20°C

(*e*) Responses to Ethylene

Exposure to ethylene at 1 ppm or higher stimulates respiration and ethylene production rates, but it does not affect their quality attributes. Pomegranates do not ripen after harvest and must be picked fully ripe to ensure the best eating quality.

(*f*) Responses to Controlled Atmospheres (CA)

Very few studies of the responses of pomegranates to CA have been reported. Storage in 2 per cent O_2 reduces chilling injury if pomegranates are kept below 5°C. The pomegranates can be stored successfully at 6°C in 3 per cent O_2 + 6 per cent CO_2 atmosphere for 6 months, whereas at

7°C a combination of 5 per cent O_2 + 15 per cent CO_2 was found to be effective in decay control and scald prevention for up to 5 months.

CA Effects

Air **5 per cent O_2 + 15 per cent CO_2**

3.5 Month Storage

For removal of pomegranate arils for further its storage and processing pomegranate aril extractor developed at CIPHET, Ludhiana/Abohar is found to extract and separate more than 90 per cent of arils.

8. Postharvest Disorders, Diseases and their Control

Physiological Disorders

Chilling Injury

External symptoms include brown discoloration of the skin and increased susceptibility to decay. Internal symptoms include a pale color of the arils (pulp around the seeds) and brown discoloration of the white segments separating the arils. Chilling injury occurs if pomegranates are exposed for longer than one month at temperatures between their freezing point 3°C and 5°C or longer than two months at 5°C.

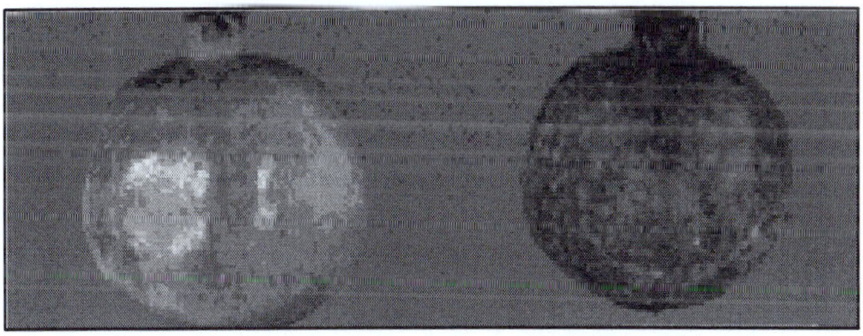

6 weeks at 5°C + 3 days at 20°C **8 weeks at 1°C + 3 days at 20°C**

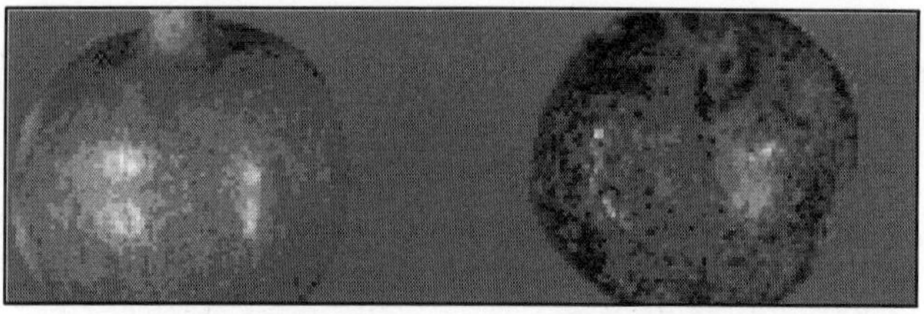

8 weeks at 10°C + 3 days at 20°C

8 weeks at 2.2°C + 3 days at 20°C

Chilling Injury

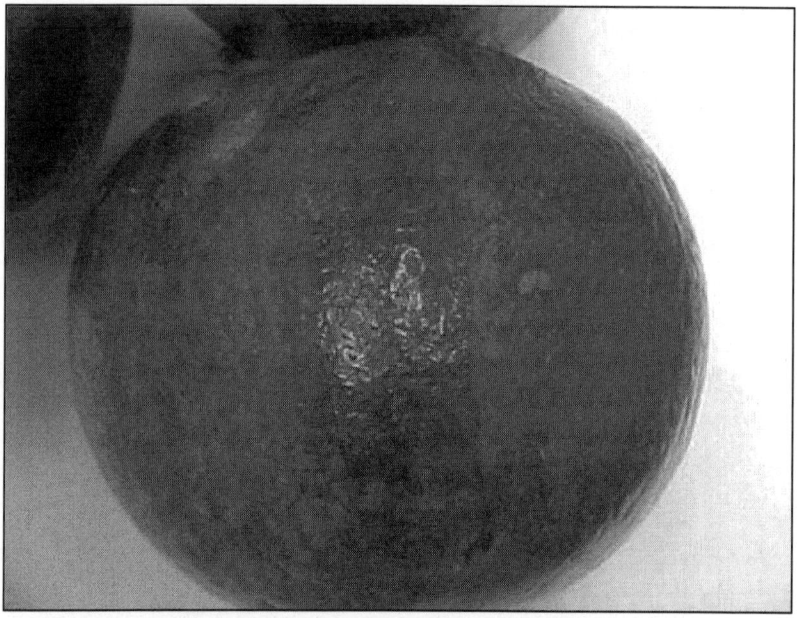

Sun Scald

Sunburn

Husk Scald

Brown discoloration of the husk (without any internal symptoms on the arils or surrounding tissues) that occurs during storage for more than 3 months at 7°C or lower temperatures.

Pathological Disorders

Heart Rot

This may be caused by *Aspergillus* spp. and *Alternaria* spp. Affected fruit show a slightly abnormal skin color, and internally a mass of blackened arils. The disease develops while the fruit is on the tree. Affected pomegranates can be detected and removed by sorters in the pack house.

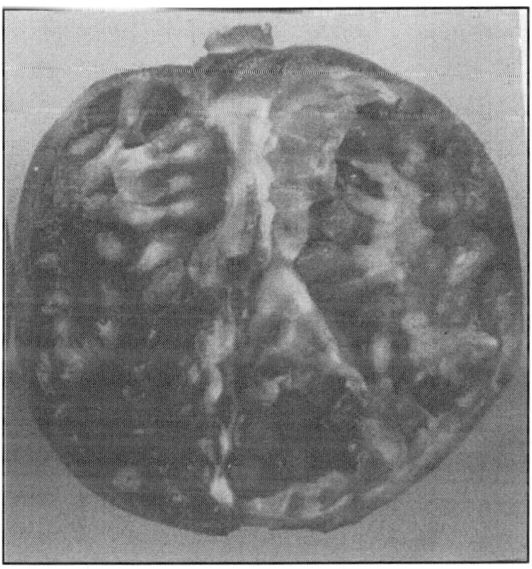

10
Apricot

Botanical Name: ***Prunus armeniaca*** **L.**

Family: Rosaeceae

1. Introduction

Apricot is a delicious fruit having characteristic, pleasing flavour. Mostly it is consumed fresh. Cultivated apricot has its origin in North Eastern China, whereas wild apricot, popularly known as Zardalu is indigenous to India. It grows wild in the hills of Shimla and Kinnaur districts of Himachal Pradesh. Fruit is delicious, rich in vitamin A and

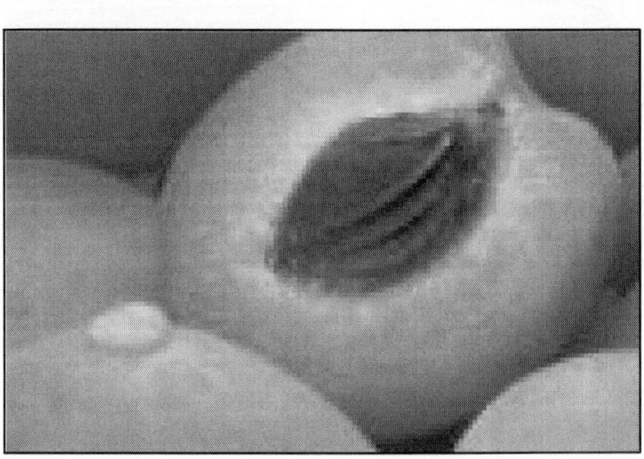

contains higher carbohydrate, proteins, phosphorus and niacin than many other common fruits. The kernel which is sweet or bitter depends upon the variety. Apricot is grown commercially in the hills of Himachal Pradesh, J and K, Uttar Pradesh and north east hills. Some drying type apricots are being grown in the areas of Kinnaur, Lahaul Sipity, in H.P and Ladakh of J and K.

2. Composition (per 100g of Edible Portion)

Constituent	Composition
Calories	48
Protein (g)	1.4
Carbohydrates	11.12
Fat (g)	0.39
Vitamin A (IU)	2612
Potassium (mg)	296
Sodium (mg)	1
Iron (mg)	0.54
Ascorbic acid (mg/100g)	10
Fiber (g)	0.6

3. Suitable Varieties

For H.P.

Mid Hill: New Castle, Early Shipley, Shakarpara,

High HillS: Kaisha, Nugget, Royal, Suffaida, Charmagz, Nari

U.P

Charmagz, Kaisha, Turkey, Chaubatia Alankar, Chaubatia Kesri

J&K

Halman, Rakchakarpa, Tokpopa, Margulam, Narma, Khanta, Turkey, Australia, Rogan, Shakarpara, Baiti, Beladi, Farmingdale, Alfred

5. Maturity Indices

The harvest date is determined when skin ground color changes from green to yellow. The exact yellowish-green color depends on the cultivar. Apricots should be picked when still firm because of their high bruising susceptibility when soft. Most apricot cultivars soften very fast making them very susceptible to bruising and subsequent decay.

Maturity and Ripeness Stages

6. Quality Indices

Fruit size, shape, and free from defects and decay. High consumer acceptance is attributed to with high (>10 per cent) soluble solids content (SSC) and moderate acidity (0.7-1.0 per cent). Apricots with 100-200 gm force flesh firmness are considered "ready to eat". Apricot cultivars have a rapid rate of fruit softening (200gm force per day at 20°C).

7. Postharvest Management

Harvesting and Packaging

Apricot fruit attains a characteristic flavor if it is allowed to ripen on the tree. As the fruit is left on the tree longer, the soluble solids content increases, firmness decreases, and consumer acceptance increases. Since apricots do not mature uniformily, selective harvesting involving two or more pickings. Selected trees with the highest preportion of mature fruit can be harvested completely. Before packing fruits are graded according to their size. Fruits are packed in CFB Boxes. Each box is lined inside with news paper sheets keeping the margins for overhanging the flaps. Fruits are arranged in layers and top layer is covered with paper by bringing together overhanging flaps. Small sized CFB boxes are also used for packaging.

Storage

(*a*) Optimum Temperature

-0.5 to 0°C is recommended. Susceptibility of cultivars to freezing injury depends on SSC, which may vary from 10-14 per cent. Highest freezing point is –1.0°C.

(*b*) Optimum Relative Humidity

90 to 95 per cent

(*c*) Rates of Respiration/CO_2 Production

Temperature °C	*ml CO_2/kg hr*
0	2-4
10	6-10
20	15-25

To calculate heat production, multiply ml CO_2/kg hr by 440 to get BTU/ton/day or by 122 to get kcal/metric ton/day.

(*d*) Rates of Ethylene Production

Ethylene production rates increase with ripening and storage temperature [<0.1 µl/kg · hr at 0°C to 4-6 µl/kg · hr at 20°C for firm-ripe apricots and higher for soft-ripe apricots].

(*e*) Responses to Ethylene

Exposure to ethylene hastens ripening (as indicated by softening and color changes from green to yellow). Also, ethylene may encourage growth of decay-causing fungi.

(*f*) Responses to Controlled Atmosphere (CA)

The major benefits of CA during storage/transport are to retain fruit firmness and ground color. CA conditions of 2-3 per cent O_2 + 2-3 per cent CO_2 are suggested for moderate benefits; extent of benefits depends on cultivar. Exposure to <1 per cent O_2 may result in development of off-flavors and > 5 per cent CO_2 can cause flesh browning and loss of flavor.

7. Disorders, Diseases and their Control

Physiological Disorders

Gel Breakdown or Chilling Injury

This physiological problem is characterized in the earlier stages by the formation of water-soaked areas that subsequently turn brown. Breakdown of tissue is sometimes accompanied by sponginess and gel formation. Fruit stored between 2.2-7.6°C have short market life and lose flavor. Market life is also related to cultivar.

Pathological Disorders

Brown Rot

It is caused by *Monilia fructicola*. It is the most important postharvest disease of apricot. Infection begins during flowering. Fruit rots may occur before harvest, but often occur postharvest. Orchard sanitation to minimize infection sources, pre-harvest fungicide application and prompt cooling after harvest are some of the control strategies.

Brown Rot

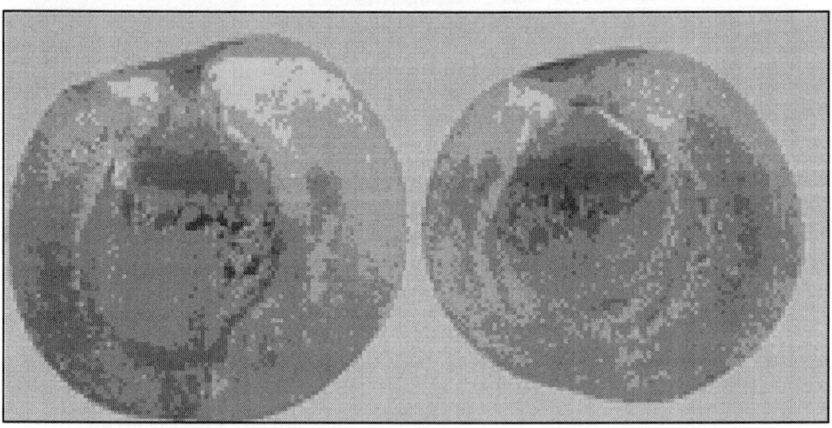

Pit Burn

Mechanical Damage

Rhizopus Rot

It is caused by *Rhizopus stolonifer*. It occurs frequently in ripe or near-ripe apricot fruits held at 20 to 25°C. Cooling the fruit and keeping them below 5°C is very effective against this fungus.

11

Banana

Botanical Name: *Musa paradisiaca*

Family: Musaceae

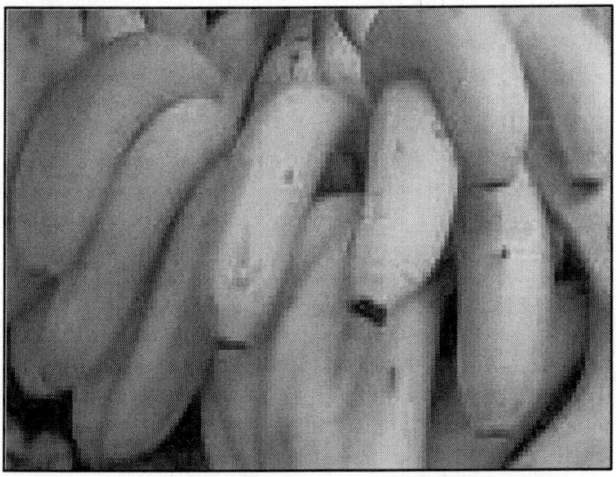

1. Introduction

Banana is one of the most important fruit crops grown with a global annual production of about 45 million metric tones. Asia contributes about 40 per cent of the banana produced in the world. India, Brazil, Phillipines, Ecuador, Indonesia China, and Thailand are the major banana producing countries.

2. Suitable Cultivars

From a genetic makeup that seems to be almost wholly derived from *M. acuminata* come the dessert bananas of world trade designated *Musa* (AAA) group indicate their triploid characters and *Acuminate* (AA) group. Gross Micheal, Lacaton, Robusta, Giant Cavendish, Gros Michel, Grand Nane, Rasthali, Poovan, Nendran, Red Banana are the major banana varieties being grown globally.

3. Composition (Per 100g Edible Portion)

Constituent	Composition
Moisture (per cent)	70
Carbohydrates (g)	27
Crude Fibre (g)	0.5
Proteins (g)	1.2
Lipids (g)	0.3
Minerals (g)	0.9
Phosphorus (mg)	29
Calcium (mg)	8
Iron (mg)	0.6
Beta Carotene (mg)	0.05
Riboflavin (mg)	0.05
Niacin (mg)	0.7
Ascorbic Acid (mg)	12
Energy, Cal	104

4. Maturity Indices

Degree of fullness of the fingers, *i.e.*, disappearance of angularity in a cross section. Bananas are harvested mature-green and ripened upon arrival at destination markets since fruits ripened on the plant often split and have poor texture.

5. Quality Indices

Maturity (the more mature the better the quality when ripe); finger length (depending on intended use and demand for various sizes); free from defects, such as insect injury, physical damage, scars, and decay.

As bananas ripen their starch content is converted into sugars (increased sweetness). Other constituents that influence flavor include acids and volatiles.

6. Postharvest Management

Harvesting and Packaging

The fruit is harvested when the ridges on the surface of the skin changes from angular to round, *i.e.* after the attainment of three fourth full stage. Dwarf bananas are ready for harvest in 11-14 months after planting. While tall varieties take about 14-16 months to harvest. The stems are usually harvested by cutting the banana plant while taking precaution to prevent damage to the fruits. The practice of handling should be to cut the individual hand of the fruits from the stalk; wash them briefly to prevent staining by the extruded latex, treat the cut surfaces with fungicide and pack the hands in cartons (CFB). However, in India banana bunchs in bulk are transported from the place of production to distant parts of the country. Hence bruising during transport is main cause of quality deterioration.

Storage

For safely transportation to ripening chambers banana hands (combs) can be removed by using banana comb cutter developed at CIPHET, Ludhiana.

(*a*) Optimum Temperature

13-14°C for storage and transport

15-20°C for ripen

(*b*) Optimum Relative Humidity

90-95 per cent

(*c*) Rates of Respiration/CO_2 Production

Temperature	13°C	15°C	18°C	20°C
ml CO_2/kg hr[1, 2]	10–30	12–40	15–60	20–70

[1] Low end for mature-green bananas and high end for ripening bananas

[2] To calculate heat production multiply ml CO_2/kg h by 122 to get kcal/metric ton/day.

(*d*) Rates of Ethylene Production

Temperature	13°C	15°C	18°C	20°C
ml C_2H_4/kg hr[1]	0.1–2	0.2–5	0.2–8	0.3–10

[1] Low end for mature-green bananas and high end for ripening bananas

(*e*) Responses to Ethylene

Most commercial cultivars of bananas require exposure to 100-150 ppm ethylene 24-48 hours at 15-20°C and 90-95 per cent relative humidity to induce uniform ripening. Carbon dioxide concentration should be kept below 1 per cent to avoid its effect on delaying ethylene action. Use of a forced-air system in ripening rooms assures more uniform cooling or warming of bananas as needed and more uniform ethylene concentration throughout the ripening.

Ethylene Effects

Control **C_2H_4 Treated**

After 7 days at 20°C

(f) Responses to Controlled Atmospheres (CA)

- 2-5 per cent O_2 and 2-5 per cent CO_2
- CA delays ripening and reduces respiration and ethylene production rates.
- Postharvest life potential of mature-green bananas: 2-4 weeks in air and 4-6 weeks in CA at 14°C
- Exposure to<1 per cent O_2 and/or >7 per cent CO_2 may cause undesirable texture and flavor.
- Use of CA during transport to delay ripening has facilitated picking bananas at the full mature stage.

7. Postharvest Diseases, Disorders and their Control

Physiological and Physical Disorders

Cold Storage Injury

Symptoms include surface discoloration, dull or smokey anal color, subepidermal tissues reveal dark-brown streaks, failure to ripen, and, in severe cases, flesh browning. Chilling injury results from exposing bananas to temperatures below 13°C for a few hours to a few days, depending on cultivar, maturity, and temperature. For example, moderate chilling injury will result from exposing mature-green bananas to one hour at 10°C, 5 hours at 11.7°C, 24 hours at 12.2°C, or 72 hours at 12.8°C. Chilled fruits are more sensitive to mechanical injury.

Skin Abrasions

Abrasions result from skin scuffing against other fruits or surfaces of handling equipment. When exposed to low (<90 per cent) relative humidity conditions, water loss from scuffed areas is accelerated and their color turns brown to black.

Impact Bruising

Dropping of bananas may induce browning of the flesh without damage to the skin.

Pathological Disorders

Crown Rot

This disease is caused by one or more of the following fungi: *Thielaviopsis paradoxa, Lasiodiplodia theobromae, Colletotrichum musae, Deightoniella torulosa,* and *Fusarium roseum*–which attack the cut surface of the hands. From the rotting hand tissue the fungi grow into the finger neck and with time, down into the fruit.

Anthracnose

It is caused by *Colletrichum musae,* becomes evident as the bananas ripen, especially in wounds and skin splits.

Stem-end Rot

It is caused by *Lasiodiplodia theobromae* and/or *Thielaviopsis paradoxa,* which enter through the cut stem or hand. The invaded flesh becomes soft and water-soaked.

Cigar-end Rot

It is caused by *Verticillium theobromae* and/or *Trachysphaera fructigena.*

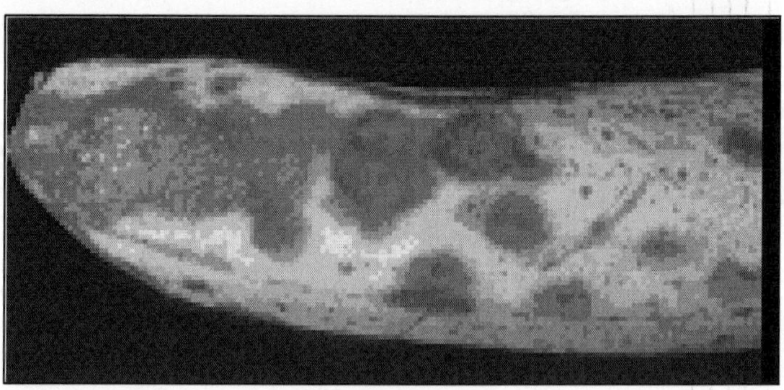

Anthracnose

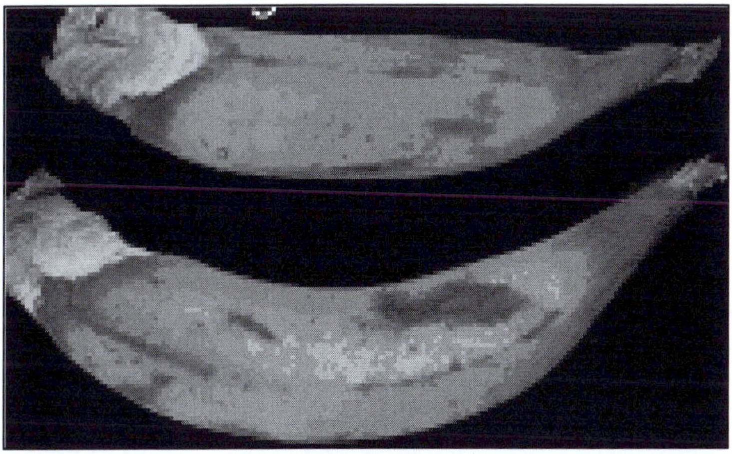

Cigar End Rot

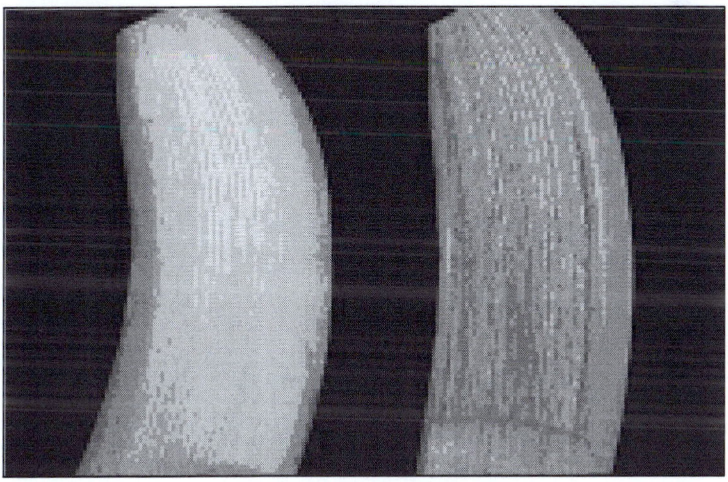

Control Chilling Injury

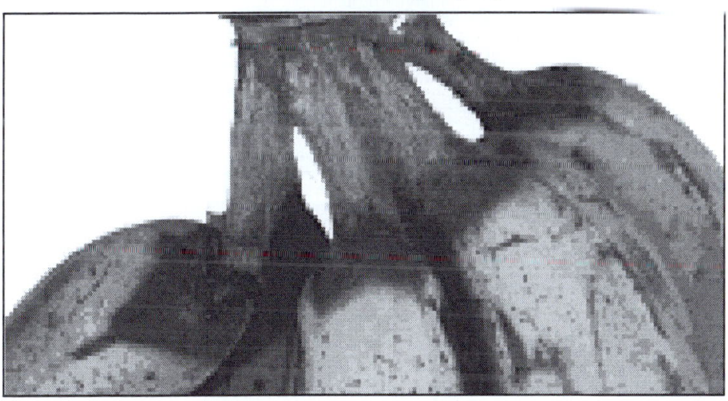

Crown Rot Chilling Injury

Chilling Injury

The rotted portion of the banana finger is dry and tends to adhere to fruits (appears similar to the ash of a cigar).

Control Strategies

Minimizing bruising; prompt cooling to 14°C; proper sanitation of handling facilities; hot water treatments [such as 5 minutes in 50°C water] and/or fungicide (such as Imazalil) treatment to control crown rot.

12

Guava

Botanical Name: ***Psidium guajava*** **L**

Family: Myrtaceae

1. Introduction

Guava is an important tropical fruit and claims superiority over other fruits by virtue of its commercial and nutritional values. It is particularly

rich in vitamin C 200–400 mg per 100 gm fresh weight. Guava is considered a common man's fruit and is rightly called the "apple of the tropics". Guava is cultivated or grown wild throughout the tropical and subtropical regions of the world. It is considered to be native of tropical America.

2. Composition (Per 100g Edible Portion)

Constituent	*Composition*
Moisture (per cent)	83.3
Dry matter (g)	16.7
Ash (g)	0.66
Crude fat (g)	0.36
Crude protein (g)	1.06
Carbohydrates (g)	14.5
Reducing (g)	4.0
Nonreducing (g)	2.8
Total sugars (g)	6.8
Calcium (mg)	0.01
Phosphorus (mg)	0.04
Iron (mg)	1.0
Vit. B 1(mg)	30.00
Riboflavin (mg)	30.0
Vitamin C (mg)	260
Crude fiber (g)	3.8

3. Suitable Varieties

Allahabad Safeda, L-49, Red Fleshed, Pant Prabhat, Pear Shaped, Apple colour

4. Maturity Indices

Guava fruits are picked at the mature-green stage (color change from dark- to light-green) in some regions where consumers eat them at that stage. In areas where consumers prefer ripe guava, the fruits are picked at the firm yellow to half-ripe (softer) stage for long-distance transport or at the fully ripe (yellow and soft) stage for local markets.

5. Quality Indices

Color is a good indicator of ripeness stage; size and shape are also important quality criteria; free from defects, insects, and decay; firmness and extent of gritty texture due to the presence of stone cells (sclereids); flesh color depends on cultivar and can be white, yellow, pink, or red; amount of seeds in the flesh (the fewer the better); aroma intensity; soluble solids and acidity.

6. Postharvest Management

Harvesting

Guava fruits are most commonly harvested by hands. Firm, yellow to half yellow, mature fruits should be harvested. Fruits for transport to distant places should be harvested mature green. Guava fruits harvested in field containers and put into boxes or bamboo basket before transport to the market or processing unit. For transport to the long distances, cushioning material such as dry grass, paddy straw, dry crushed leaves etc. should be used. Good ventilation is important to prevent the built up of heat and microbial spoilage.

Storage

(*a*) Optimum Temperature

8-10°C for mature-green and partially ripe guavas (storage potential = 2-3 weeks)

5-8°C for fully ripe guavas (storage potential = 1 week)

(*b*) Optimum Relative Humidity

90-95 per cent

(*c*) Rates of Respiration/CO_2 Production

Temperature	ml CO_2/kg hr
10°C	4-30
20°C	10-70

To calculate heat production multiply ml CO_2/kg hr by 122 to get kcal/metric ton/day.

(*d*) Rates of Ethylene Production

Guava is a climacteric fruit. Rates of respiration and ethylene production depend upon cultivar and maturity/ripeness stage. Ethylene production at 20°C ranges from 1 to 20 µl/kg•hr.

(*e*) Responses to Ethylene

Ethylene at 100 ppm for 1-2 days can accelerate ripening of mature-green guavas to full-yellow stage at 15-20°C and 90-95 per cent relative humidity. This treatment results in more uniform ripening, which is more important for guavas destined for processing. Immature-green guavas do not ripen properly and develop 'gummy' texture.

(*f*) Responses to Controlled Atmospheres (CA)

2-5 per cent oxygen levels may delay ripening of mature-green and partially-ripe guavas kept at 10°C.

7. Postharvest Disorders, Diseases and their Control

Physiological and Physical Disorders

Chilling Injury

Symptoms include failure of mature-green or partially-ripe guavas to ripen, browning of the flesh and, in severe cases, the skin, and increased decay incidence and severity upon transfer to higher temperatures. Fully-ripe guavas are less sensitive to chilling injury than mature-green guavas and may be kept for up to a week at 5°C without exhibiting chilling injury symptoms.

External (Skin) and Internal (Flesh) Browning

Guavas are sensitive to physical damage during harvesting and handling all the way to the consumer. Symptoms include skin abrasions and browning of bruised areas.

Sun Scald

Guavas exposed to direct sunlight may be scalded. Some times paper bags are used to cover guava fruits and protect them from solar radiation and insect infestation while on the tree.

Pathological Disorders

Most of the postharvest disease problems begin in the orchard as latent infection in developing fruits. Diseases include anthracnose (caused by *Colletrotrichum gloeosporioides* and associated species), aspergillus rot (caused by *Aspergillus niger*), mucor rot (caused by *Mucor hiemalis*), phomopsis rot (caused by *Phomopsis destructum*), and rhizopus rot (caused by *Rhizopus stolonifer*).

Disease control strategies include good orchard sanitation, effective preharvest management to reduce infection, careful handling to reduce physical damage, prompt cooling to 10°C and subsequent maintenance of that temperature throughout the handling system.

Insect Control

Guavas are a preferred host for fruit flies and must be treated for disinfestation. One of the insect control treatments is immersion in 46°C-water for 35 minutes or exposed to hot air at 48°C for 60 minutes. Another potential insect control treatment is irradiation at 0.15-0.30 kGy.

13

Sweet Orange

Botanical Name: *Citrus chinensis*

Family : Rutaceae

1. Introduction

Sweet orange are the second largest fruits cultivated in the country. Andhra Pradesh, Maharashtra, Karnataka, Punjab, Haryana, and Rajasthan are main sweet orange growing states. Maximum area under sweet orange is in Andhra Pradesh followed by Maharashtra, and Karnataka. Well marked belt of sweet orange cultivation in the country are Abohar, Fazilka, Ferozpur, Faridkot, and Hoshiarpur in Punjab: Hissar in Haryana: Ganganagar in Rajasthan: Marathwara region of western Maharashtra, : Ahmed Nagar, Nasik and Pune in Maharashtra and Anandpur, Kodur, Cuddapa, Negonda, Chittoor districts of Andhra Pradesh.

2. Composition

Constituent	*Composition*
Moisture (per cent)	88.4
Crude fat (g)	0.3
Crude protein (g)	0.8
Carbohydrates (g)	9.3
Minerals(g)	0.7
Total sugars (g)	
Calcium (mg)	40
Phosphorus (mg)	30
Iron (mg)	0.7
Vitamin C (mg)	50
Crude fiber (g)	0.5
Energy K cal	43

3. Suitable Varieties

Jaffa, Hamlin, Pineapple, Pineapple Blood Tred, Mosambi and Satguddi

4. Maturity Indices

Color (yellow, orange, and/or red) on 75 per cent of fruit surface and soluble solids/acid ratio of 6.5 or higher.

5. Quality Indices

Color intensity and uniformity; size; shape; firmness; free from decay; and free from defects including insect damage, and scars. Flavor depends upon soluble solids/acid ratio and absence of off-flavors.

6. Postharvest Management

Harvesting

Oranges should be preferrably harvested by snub-nosed clippers. The fruits should be harvested in dry conditions. Fruits must be carefully handled during and after picking.

Storage

After harvesting to clean the fruit from dirt, soil, insect excreta and sap fruit washing machines developed at PAU, Ludhiana is found very effective.

(*a*) Optimum Temperature

5-8°C for 2 to 6 weeks, depending on cultivar, maturity-ripeness stage at harvest, and decay control treatments used.

(*b*) Optimum Relative Humidity

90-95 per cent

(*c*) Rates of Respiration/CO_2 production

Temperature	5°C	10°C	15°	20°
ml CO_2/kg hr	2-4	3-5	6-10	10-15

To calculate heat production multiply ml CO_2/kg hr by 440 to get Btu/ton/day or by 122 to get kcal/metric ton/day.

(*d*) Rates of Ethylene Production

< 0.1 µl/kg·hr at 20°C

(*e*) Responses to Ethylene

Mandarins and tangerines can be degreened by exposure to 1-10 ppm ethylene for 1-3 days at 20 to 25°C

Removal of ethylene from transport vehicles and storage facilities for citrus fruits can help reduce decay incidence.

(*f*) Responses to Controlled Atmospheres (CA)

A combination of 5-10 per cent O_2 and 0-5 per cent CO_2 can delay color changes from green to yellow and other symptoms of senescence, but it is not very effective in decay control. Mandarins do not tolerate exposure to fungistatic CO_2 levels (10-15 per cent). Commercial use of CA is very limited.

For short storage CIPHET, Ludhaina evaporative cooled storage structure having capacity of 5–7 tonnes can be effectively used.

7. Postharvest Diseases, Disorders and their Control

Physiological Disorders

Chilling Injury

Symptoms include pitting and brown discoloration followed by increased susceptibility to decay. Severity increases with longer exposures to lower temperature below 5°C.

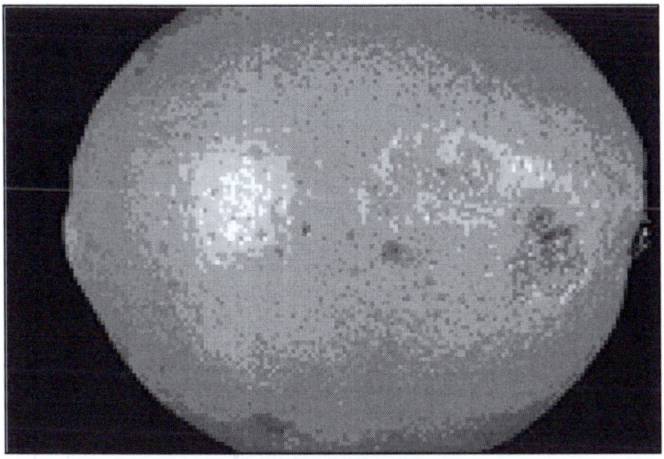

Chilling Injury

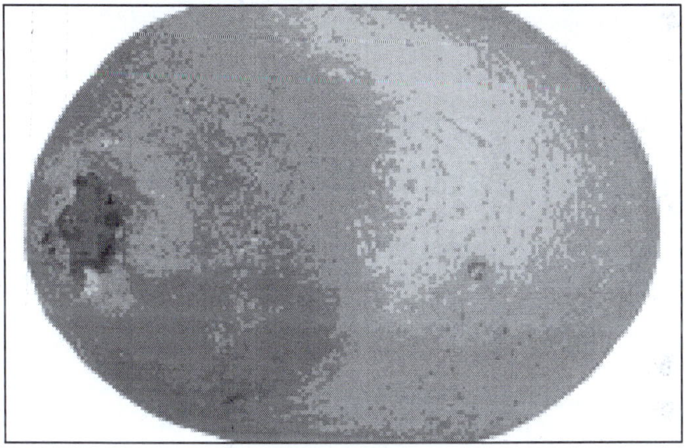

Anthracnose

Oil Spotting (Oleocellosis)

Harvesting and handling turgid citrus fruits can result in breaking of oil cells and release of oil that damages surrounding tissues.

Ageing

Symptoms include shriveling and peel injury around the stem end.

Pathological Disorders

Major Diseases

- ✰ Green mold *(Penicillium digitatum)*
- ✰ Blue mold *(Penicillium italicum)*
- ✰ Phomopsis stem-end rot *(Phomopsis citri)*
- ✰ Stem end rot *(Lasiodiplodia theobromae)*

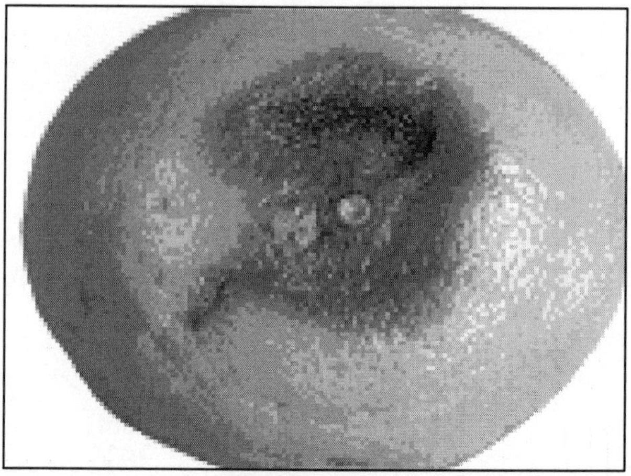

Alternaria Rot

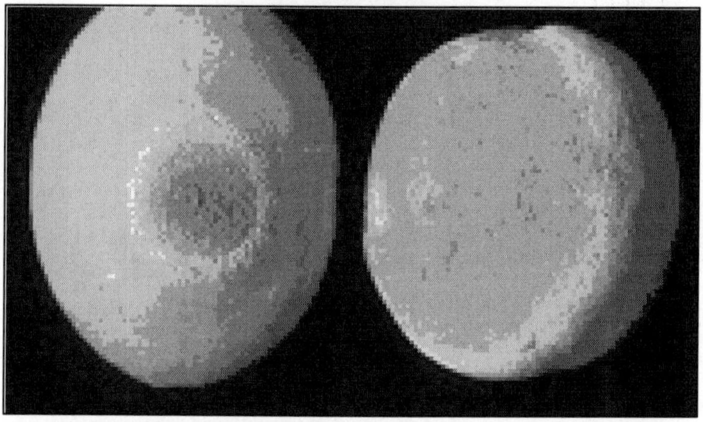

Penicillium italicum ***Penicillium digitatum***

Blue Green Rot

B. Cinera

Botrytis Rot

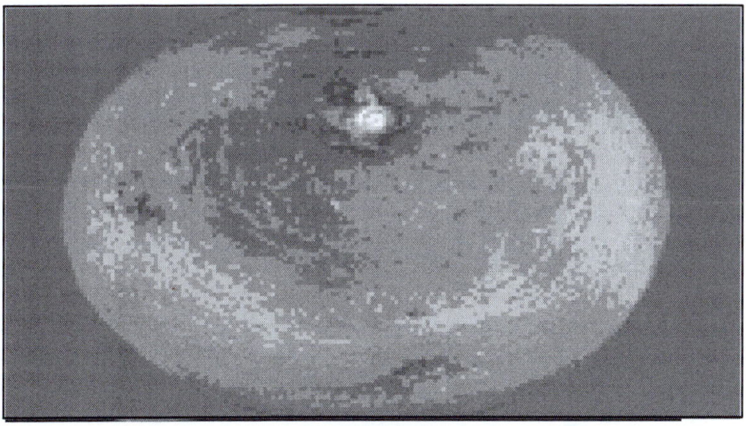

Phomopsis citri
Phomopsis

- Brown rot *(Phytophthora citrophthora)*
- Anthracnose *(Colletotrichum gloeosporioides)*

Control Strategies

Reduce the Pathogen Population in the Environment

- Effective preharvest disease control.
- Use of chlorine in wash water.
- Heat treatments.
- Effective sanitation procedures

Maintain Fruit Resistance to Infection

- Minimize mechanical injuries.
- Use proper ranges of temperatures and relative humidity throughout postharvest handling.
- Use postharvest fungicides and/or biological antagonists.
- Avoid exposure to ethylene.

14

Mandarin Orange

Botanical Name: *Citrus reticulata*

Family: Rutaceae

1. Introduction

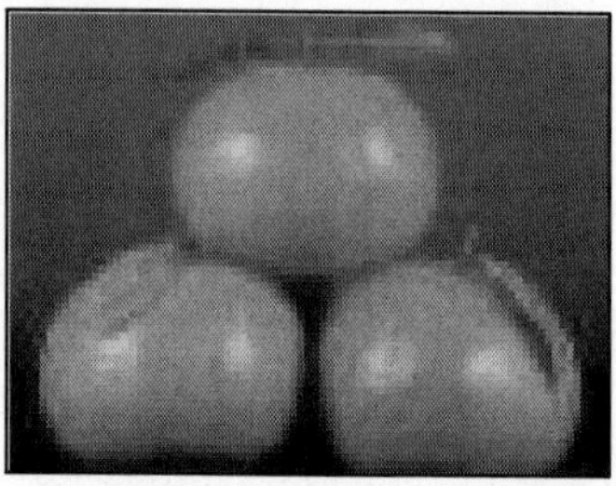

Mandarin oranage is most common among citrus fruits grown in India. It occupies nearly 50 per cent of the total citrus area in India. Though, it is grown in every state, certain belts/pockets have emerged as the leading producers. Nagpur Santra (mandarin) is mainly grown in Satpura hills (Vidarbha region) of central India. Hilly slopes of Darjeeling (West Bengal) and Coorg (Karnataka) are other major belts of mandarin production. In North-western India, Kinnow mandarin is being grown satisfactorily in Punjab, Rajasthan, Haryana, Himachal Pradesh, Jammu and Kashmir and Uttar Pradesh. In South India, Wynad, Nilgril, Palney and Shevroy hills are major mandarin-growing belts, while hills of Meghalaya (Khasi, Dusha, Garo, Jaintia), Mizoram, Tripura, Sikkim and Arunchal Pradesh have predominance in mandarins. In Assam, Brahmaputra valley and Dibrugarh districts are famous for mandarin production.

2. Composition

Constituent	*Composition*
Moisture (per cent)	87.6
Crude fat (per cent)	0.2
Crude protein (per cent)	0.7
Carbohydrates (per cent)	10.9
Total sugars (per cent)	
Calcium (per cent)	26
Phosphorus (per cent)	20
Iron (per cent)	0.32
Carotene (µg/100g)	1104
Vitamin C (mg/100g)	30
Crude fiber (per cent)	0.3
Energy K cal	48

3. Suitable Varieties

Coorg, Khasi, Nagpur, Kinnow

4. Maturity Indices

Soluble solids/acid ratio of 8 or higher and yellow-orange color at least on 25 per cent of the fruit surface or soluble solids/acid ratio of 10 or higher and green-yellow color on 25 per cent or greater of the fruit surface.

5. Quality Indices

Color intensity and uniformity; firmness; size; shape; smoothness; free from decay; and free from defects including physical damage (abrasions and bruising), skin blemishes and discoloration, freezing damage, chilling injury, and insect damage. Flavor quality is related to soluble solids/acid ratio and absence of off-flavor-causing compounds including fermentative metabolites

6. Postharvest Management

Harvesting

Mandarins should be preferrably harvested by snub-nosed clippers. The fruits should be harvested in dry conditions. Fruits must be carefully handled during and after picking.

Storage

(*a*) Optimum Temperature

3-8°C for up to 3 months, depending on cultivar, maturity-ripeness stage at harvest and production area.

(*b*) Optimum Relative Humidity

90-95 per cent

(*c*) Rates of Respiration/CO_2 production

Temperature	5°C	10°C	15°	20°
ml CO_2/kg hr	2–4	3–5	6–12	11–17

To calculate heat production multiply ml CO_2/kg hr by 122 to get kcal/metric ton/day.

(*d*) Rates of Ethylene Production

< 0.1 µl/kg·hr at 20°C

(*e*) Responses to Ethylene

Exposure to 1-10ppm ethylene for 1-3 days at 20-30°C may be used for degreening oranges. This treatment does not influence the internal quality (including soluble solids/acid ratio) and may accelerate deterioration and decay incidence.

(*f*) Responses to Controlled Atmospheres (CA)

A combination of 5-10 per cent O_2 and 0-5 per cent CO_2 can be useful for delaying senescence and for firmness retention but does not have a significant effect on decay incidence and severity, which is the limiting factor to long-term storage of oranges. Fungistatic levels (10-15 per cent) of CO_2 are not used because they may result in off-flavors due to accumulation of fermentative metabolites. Commercial use of CA on oranges during storage and transport is very limited.

(*g*) Effect of Shrink Wrapping and Storage Life

Results of various experiments conducted at CIPHET, Abohar reveals that film wrapping of mandarins could be an alternative for controlling water loss, spread of decay and retention of fruit shape without adverse effect on flavour and colour development. Tray or individual cover wrapping with 15µ heat shrinkable film and storage at cold storage is most effective in reducing the weight loss, decay and maintain the quality in terms of TSS, acidity, vitamin C colout, aroma and flavour of the juice.

7. Postharvest Disorders, Diseases and their Control

Physiological Disorder

Chilling Injury/Cold Storage Chilling

Symptoms include pitting, brown staining, and increased decay incidence. Minimum safe temperature depends on cultivar, production area, and maturity-ripeness stage at harvest. Severity of symptoms can be reduced if water loss is minimized (by waxing or film wrapping) and if decay-causing fungi are controlled (by use of fungicides and/or biological antagonists).

Stem-end Rind Breakdown

Symptoms include shriveling and peel injury around the stem due to ageing.

Rind Staining

This disorder results from overmaturity at harvest. It can be reduced by preharvest application of gibberellic acid that delays senescence.

Oil Spotting (Oleocellosis)

Harvesting and handling turgid oranges can result in release of oil that damages surrounding tissues. Thus, oranges should not be harvested when fully turgid such as early in the morning and soon after rain or irrigation.

Pathological Disorders

Important Diseases

- Green mold *(Penicillium digitatum)*
- Blue mold *(Penicillium italicum)*
- Phomopsis stem-end rot *(Phomopsis citri)*
- Stem end rot *(Lasiodiplodia theobromae)*
- Brown rot *(Phytophthora citrophthora)*
- Sour rot *(Geotrichum candidum)*

0 1 2 3 4 5

Degreening

Control Strategies

- ☆ Minimizing physical damage during harvesting and handling.
- ☆ Treatment with postharvest fungicides and/or biological antagonists. Also, heat treatments may be used.
- ☆ Prompt cooling and subsequent maintenance of optimum temperature and relative humidity throughout marketing operations.
- ☆ Removal and/or exclusion of ethylene.
- ☆ Effective sanitation procedures throughout postharvest handling.

15

Passion Fruit

Botanical Name: *Passiflora edulis*

Family: Passifloraceae

1. Introduction

Passion fruit is native of Brazil. Passion fruit is native to tropical America. It is grown in practically every country with a suitable climate. There are two types, yellow and purple. In India it grows wild in the Nilgiris, Wyanad, Kodaikanal, Coorge and Malabar. Recently its cultivation has been extended to some areas in Andaman and Nicobar Islands, Nagaland, Sikim and Mizoram. The juice has excellent flavor and is commonly used for preparation of beverages, cakes, pies, and ice creams.

2. Composition (Per 100g Edible Portion)

Constituents	Composition	
	P. ediulis f. edulis	*P. edulis f. flavicarpa*
Water (per cent)	85.6	84.9
Energy (cal)	51.0	53
Protein (g)	0.4	0.7
Fat (g)	0.1	0.2
Carbohydrates		
Total (g)	13.6	13.7
Fiber (g)	0	0.2
Ash (g)	0.3	0.5
Calcium (mg)	3.6	3.8
Phosphorus (mg)	12.5	0.4
Iron (mg)	0.2	0.4
Sodium (mg)	-	-
Potassium (mg)	-	-
Vitamin A (I.U.)	717	2410
Thiamin (mg)	Trace	Trace
Riboflavin (mg)	0.1	0.1
Niacin (mg)	1.5	2.2
Ascorbic acid (mg)	30	20

3. Suitable Varieties

Purple (*P. edulis*), Yellow (*P. edulis* var. *flavicarpa*), Kaveri (Hybrid of Purple x Yellow)

4. Maturity Indices

The amount of yellow or purple color on the fruit surface is used as a maturity index for fresh market passion fruits. In some cases fruits are allowed to fall and are collected from the ground for processing into juice concentrate, jam, and other products.

5. Quality Indices

The fruit is a berry 3.5 to 7 cm wide and 4 to 12 cm long and has a moderately hard shell (3-10 mm thick), depending on cultivar. The edible portion is the fleshy, acidic pericarp together with the arils surrounding the seeds. Fruit color may be purple or yellow. Soluble solids content ranges between 14 and 18 per cent and acidity ranges from 3 to 5 per cent in the pulp. Moisture loss during ripening may be large enough to cause shriveling of passion fruits, but this does not influence the edible portion.

6. Postharvest Management

Harvesting

The fruit is generally not picked, but left on the vine until it falls for full development of the flavor. The fallen fruits are gathered every morning. The picking of matured fallen fruits from the ground at an interval of 2–7 days is the common method of harvesting.

Storage

(*a*) Optimum Temperature

7–10°C for partially-ripe fruits (potential storage life – 3–5 weeks) 5–7°C for fully ripe fruits (potential storage life = 1 week)

(*b*) Optimum Relative Humidity

90-95 per cent

(c) Rates of Respiration/CO_2 Production

Temperature	5°C	10°C	20°C
ml CO_2/kg hr	15-30	20-40	45-100

To calculate heat production multiply ml CO_2/kg hr by 122 to get kcal/metric ton/day.

(d) Rates of Ethylene Production

Passion fruits are the highest ethylene producers among all fruits with a production range of 160-370 µl/kg·hr at 20°C at their climacteric peak.

(e) Responses to Ethylene

Exposure of mature-green passion fruits to 100 ppm ethylene for 1-2 days accelerates their ripening. Once ripening begins external ethylene treatment is unnecessary because the fruits produce high ethylene concentrations.

(f) Responses to Controlled Atmospheres (CA)

Passion fruits may benefit from packaging in perforated plastic films (no or minor effect on atmospheric modification) due to reduction in water loss during handling.

7. Postharvest Disorders, Diseases and their Control

Physiological and Physical Disorders

Chilling Injury

Symptoms occur on passion fruits kept at 5°C or lower and include surface and internal discoloration, pitting, water-soaked areas, uneven ripening or failure to ripen, off-flavor development, and increased decay incidence.

Pathological Disorders

Brown Spot

Caused by *Alternaria passiflorae* and appears as circular, sunken, light-brown spots on ripening fruits. Disease incidence is most severe during warm wet periods.

Phytophthora Fruit Rot

Caused by *Phytophthora nicotianae* var. *parasitica* and appears as water-soaked, dark-green patches which dry up.

Septoria Spot

Caused by *Septoria passiflorae* which infects fruits while on the plant and results in uneven ripening.

Control Procedures

Effective vineyard sanitation, pruning and leaf thinning to allow more air and light to reach the canopy, application of preharvest fungicides, and proper management of temperature and relative humidity during postharvest handling.

16
Sweet Cherry

Botanical Name: ***Prunus avium***

Family: Rosasea

1. Introduction

Cherries are mostly eaten as dessert fruit. Deandolle reported that the cherry first grew wild in northern Persia and the Russia provinces south of the Caucasus. From there, it spread rapidly because they are attractive to birds; hence the name *Prunus avium* L. or bird cherry. In India its cultivation is confined to Kashmir, Himachal Pradesh, and hills of Uttar

Pradesh. A delicious fruit, cherry is rich in protein, sugars, and minerals, has more calorific value than apple. Due to higher return cherry is gaining popularity in temperate regions of the country.

2. Composition (Per 100g Edible Portion)

Constituent	Composition
Moisture (per cent)	83.7
Total sugars (per cent)	13.04
Sucrose (per cent)	0.10
Glucose (per cent)	4.70
Fructose (per cent)	7.24
Proteins (per cent)	0.7
Ascorbic acid (mg/100g)	13-56

3. Suitable Varieties

Jammu and Kashmir

Black Heart, Early purple Guigne Noir, Gross Lucenta, Guigne Noir Hative, Napolean, Guigne Noir gross

Himachal Pradesh

Black Tartarian, Bing, Napolean White, Sam, Sue, Stella van, Lambert, Black Republician, Pink Early, White Heart and Early Rivers.

Uttar Pradesh

Bedford prolific, Black Heart and Governers Wood.

4. Maturity Indices

Skin color and soluble solids content (SSC) are the main criteria used to judge fruit maturity. Minimum maturity is that when the entire cherry surface have a minimum of light red color and/or 14 to 16 per cent SSC, depending on the variety. The red mahogany stage is recommended for harvest for many varieties as maturity and ripeness stages of cherries.

5. Quality Indices

Taste is related to SSC, titratable acidity (TA) and the ratio of SSC/TA. Freedom from cracks, bird pecks, shriveling, decay or misshappen fruit (doubles, spurs). Green fleshy stems are often associated with freshness and quality.

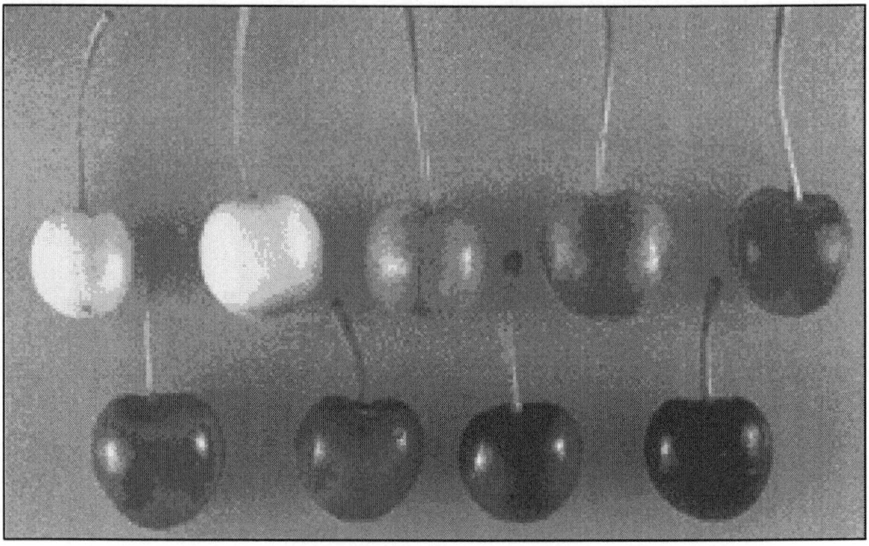

Cherry Maturity Stages

5. Postharvest Management

Harvesting

The cherries are harvested at full maturity when surface colour changes from green to red. The fruits are packed in boxes lined with paper. Generally 5 kg boxes are used for packing. For fresh market the cherries are harvested manually.

Storage

(*a*) Optimum Temperature

-0.5 ± 0.5°C

(*b*) Optimum Relative Humidity

90-95 per cent; high humidity is particularly important to maintain green stem color.

(*c*) Rates of Respiration/CO_2 Production

Temperature	0°C	5°C	10°C	20°C
ml CO_2/kg hr	3–5	5–9	15–17	22–28

To calculate heat production multiply ml CO_2/kg hr by 122 to get kcal/metric ton/day.

(*d*) Rates of Ethylene

< 1 μl/kg·hr at 20°C

(*e*) Responses to Ethylene Production

Cherry response to ethylene is minimal. Ethylene does not accelerate cherry ripening.

(*f*) Response to Control Atmosphere

CA reduces respiration rate and thereby increases postharvest life. Elevated CO_2 suppresses decay development. Modified atmosphere packaging within boxes has been very successful. Successful atmospheres are generally within the following ranges:

3 to 10 per cent O_2

10 to 15 per cent CO_2

< 1 per cent O_2 can result in skin pitting and off-flavors

> 30 per cent CO_2 can result in brown skin discoloration and off-flavors.

Flavor volatiles may be reduced following several weeks of CA storage resulting in fruit of good visual quality but poor sensory quality.

7. Postharvest Disorders, Diseases and their Control

Physiological and Physical Disorders

Pitting

An indentation in the surface of the fruit caused by the collapse of cells under the skin. Thought to be result of impact injury.

Bruising

Results from compression and impact of the fruit.

Postharvest life is closely related to respiration rate. Respiration rate increases as a result of increased temperature and physical injury.

Pathological Disorders

Brown Rot

Caused by *Monilinia fruticola,* disease can begin in the orchard or postharvest. Pre and postharvest control measures are necessary.

Cherry CA Effects

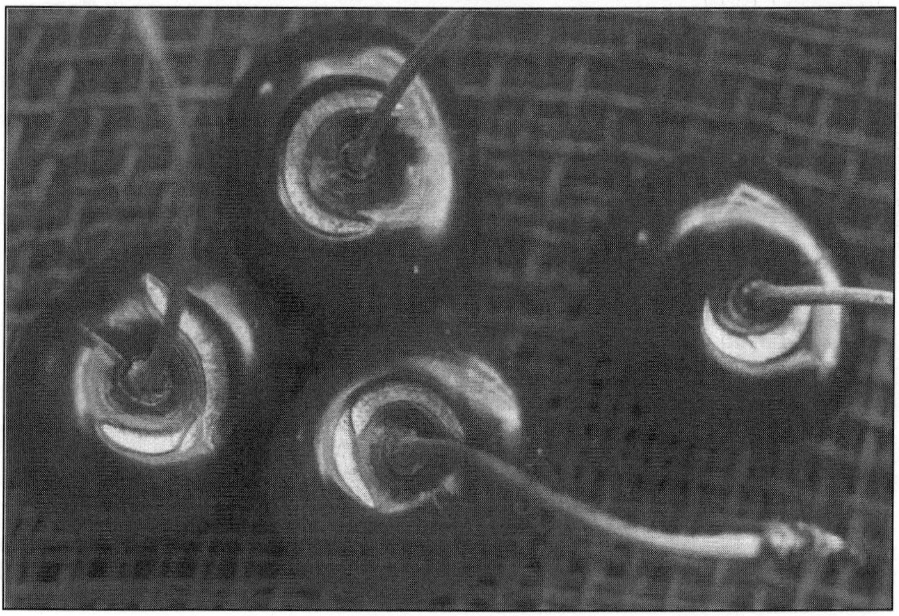

Concentric Cracking

Grey Mold

Caused by *Botrytis cinerea,* a fungus that continues to grow slowly at 0°C.

Cherry Doubles

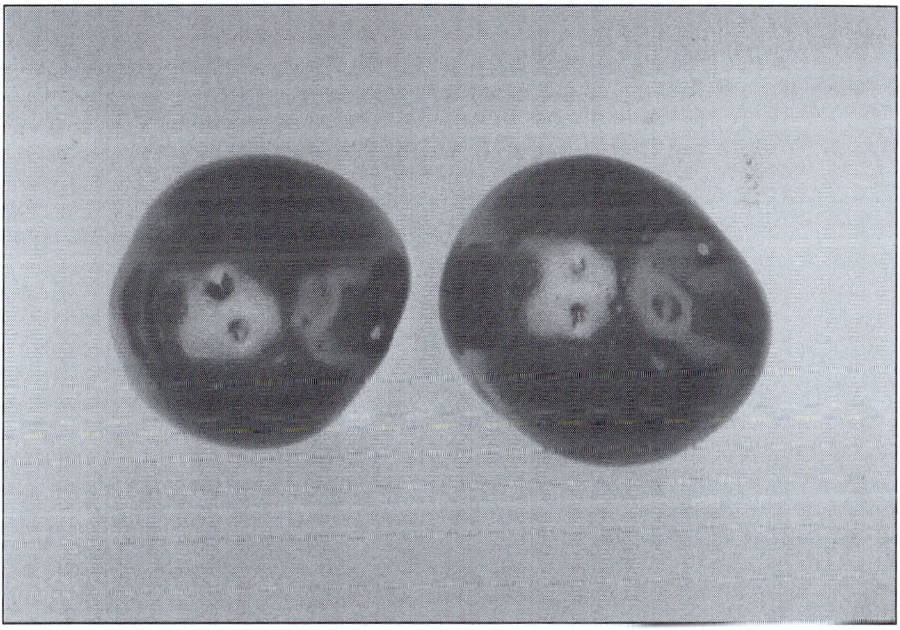

Cherry Impact Bruise

Rhizopus Rot

Caused by *Rhizopus stolonifer*, a fungus that is found in fruit exposed to temperatures of 5°C or greater.

Cherry Pitting Damage

Proper temperature management (rapid cooling to optimum storage temperature) can completely control Rhizopus Rot and significantly reduce Brown Rot and Grey Mould. Eliminating injured and diseased fruit from the packed box is important. Fungicide treatments, pre and postharvest are often beneficial.

17

Date

Botanical Name: *Phoenix dactylifera*

Family: Palmaceae

1. Introduction

Date palm is a highly nutritious fruit. It is rich in sugar, iron, potassium, calcium and nictoic acid. Date fruit can help supplement the dietary needs of desert people where very few nutritive foods are available. The leaves of the palm also have potential for use in the manufacture of paper. Long dry summer and sufficient heat unit accumulation for development and ripenning of fruit, sufficient water resources for irrigation and production technology suitable for Indian agroclimate make India quite suitable for its commercial cultivation. The extremely dry areas comprising Jaisalmer, Barmer and western parts of Bikaner and Jodhpur districts is the potential region for its cultivation. In other areas, fruits are harvested either at dang or rutab (soft translucent) stage (eastern parts of Jodhpur, Bikaner, Barmer and Western parts of Nagpur).

2. Composition (Per 100g Edible Portion)

Constituent	*Composition*	
	Dried dates	*Fresh*
Moisture (per cent)	15.3	59.2
Protein (g)	2.5	1.2
Fat (g)	0.4	0.4
Minerals (g)	2.1	1.7
Fibre (g)	3.9	3.7
Carbohydrate (g)	75.8	33.8
Energy (k cal)	317	144
Calcium (mg)	120	22
Phosphorus (mg)	50	38
Iron (mg)	7.3	0.96
Carotene (μg)	26	
Thiamine (mg)	0.01	
Riboflavin (mg)	0.02	
Niacin (mg)	0.9	
Vitamin C (mg)	3	

3. Suitable Cultivars/Varieties

Halway, Khadarawy, Shamran (Sayer), Medjool, Barhee, Zahidi, Khalas

Cultivars

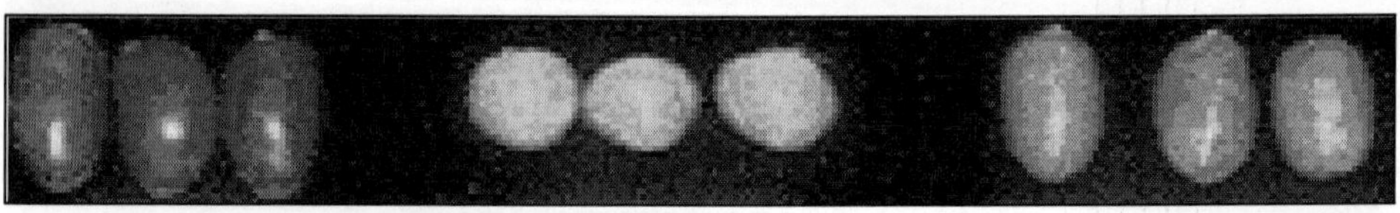

Abada **Barhee** **Deglet Noor**

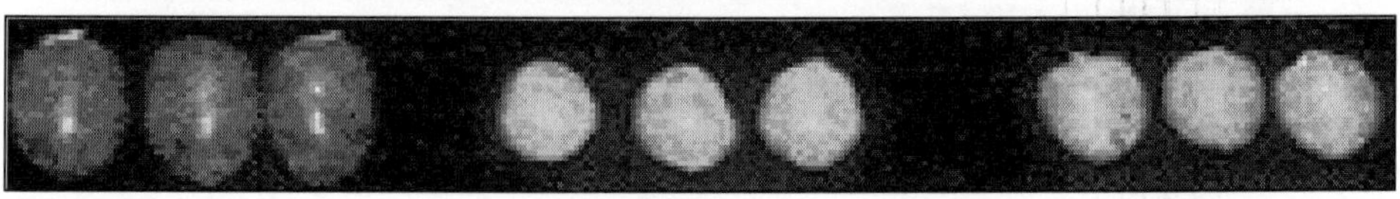

Halway **Honey** **Khadrawy**

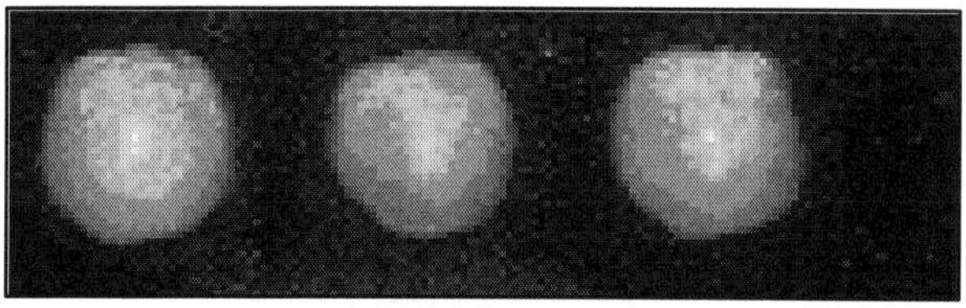

Medjool

4. Maturity Indices

A small quantity of dates is harvested at the "Khalal" stage (partially-ripe) when they are yellow or red (depending on cultivar), but many consumers find them astringent (high tannin content). Most dates are harvested at the fully-ripe "Rutab" and "Tamar" stages, when they have much greater levels of sugars, lower moisture and tannin content, and softer than the "Khalal" stage dates.

Ripening

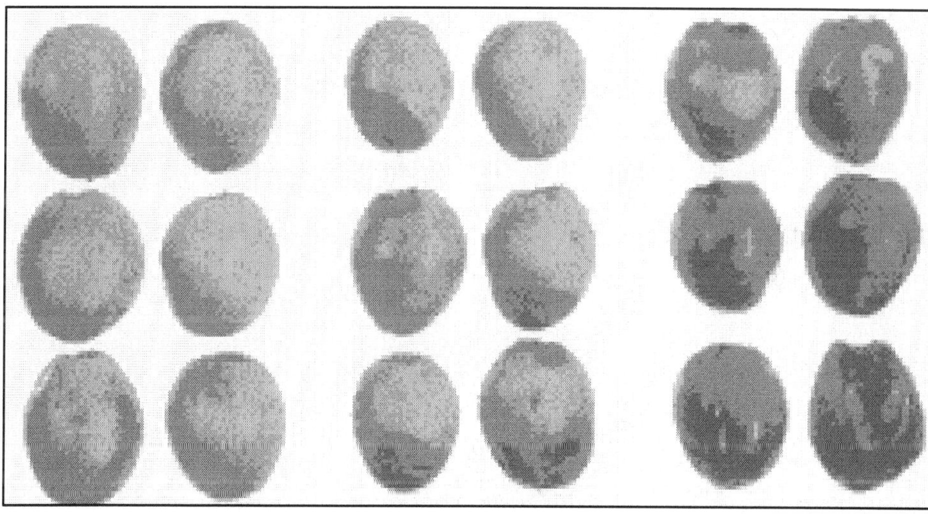

Khalal **Partially-rutab** **Rutab**

5. Quality Indices

Fruit size, color, texture, cleanliness, and free from defects (such as sunburn, insect damage, sugar migration to fruit surface, and fermentation) and decay-causing pathogens.

Sweetness. Sucrose is the main sugar in some cultivars while reducing sugars are predominant in others; total sugars represent about 50 per cent (fresh wt basis) or 75 per cent (dry wt basis).

Quality

6. Postharvest Management

Harvesting

Harvesting of the date fruits at doka stage is done by cutting the whole bunch and plucking of the berries from the bunch. It should be done at a stage when more than 80 per cent fruits in a bunch attain desirable quality. When fruits are harvested at dang stage for soft dates, only those fruits which have attained $1/3^{rd}$ ½ dang are picked and thus harvesting is done of selected fruits. For chuahara making fruits are to be harvested at full doka stage.

Storage

(*a*) Optimum Temperature

0°C for 6-12 months, depending on cultivar (semi-soft dates, such as "Deglet Noor" and Halway", have longer storage-life than soft dates, such as "Medjool" and "Barhi".

(*b*) Optimum Relative Humidity

70-75 per cent; at higher relative humidities, dates will absorb moisture from the room air unless they are packaged in moisture-proof containers.

(*c*) Rates of Respiration

<5ml CO_2/kg·hr for "Khalal" stage dates; <1ml/kg · hr for "Rutab" and "Tamar" stage dates kept at 20°C. The rates increase with the moisture content increases.

To calculate heat production, multiply ml CO_2/kg·hr by 122 to get kcal/metric ton/day.

(*d*) Rates of Ethylene Production

<0.1 μl/kg·hr for "Khalal" stage dates; none for "Rutab" and "Tamar" stage dates kept at 20° C.

(*e*) Responses to Ethylene

Ripe dates are not influenced by ethylene but can readily absorb the aroma of other products. Thus, dates should not be stored with garlic, onion, potato, or other commodities with strong order.

(*f*) Responses to Controlled Atmosphere (CA)

Packaging in nitrogen (to exclude oxygen) reduces darkening of dates and prevents insect infestation.

7. Postharvest Diseases, Disorders and their Control

Physiological and Physical Disorders

Darkening

Both enzymatic and non-enzymatic browning occur in dates and increases with higher moisture content and higher temperatures. Enzymatic browning can be inhibited at low oxygen concentrations.

Souring

Yeasty fermentation results in souring of dates with moisture content above 25 per cent.

Sugar Spotting

Crystallization of sugars below the skin and in the flesh of soft date cultivars. Although it does not influence taste it alters fruit texture and appearance. Storage at recommended temperatures minimizes this disorder, which occurs mainly in cultivars in which glucose and fructose are the main sugars.

Pathological Disorders

Microbial spoilage can be caused by yeasts (most important), molds and bacteria. Yeast *species of Zygosaccharomyces* are more tolerant of high sugar content than others found in dates. Yeast-infected dates develop an alcoholic odor (become fermented). Acetobacter bacteria may convert the alcohol into acetic acid (vinegar). Fungi (Aspergillus, Alternaria, and Penicillium spp) may grow on high-moisture dates, especially when harvested following rain or high humidity period.

Control Strategies

- ✰ Dry the dates to 20 per cent moisture or lower to greatly reduce incidence of molds and yeasts.
- ✰ Maintain recommended temperature and relative humidity ranges throughout the handling system.
- ✰ Avoid temperature fluctuations to prevent moisture condensation on dates, which may encourage growth of decay–causing microorganisms.
- ✰ Use adequate sanitation procedures in the pack house and storage rooms.

Insect Infestations

- ✰ Dates can be infested with some of the stored-products insects and must be fumigated with an approved fumigant for disinfestation followed by packaging in insect-proof containers.
- ✰ "Organic" dates may be treated with 100 per cent carbon dioxide for 1-2 days since chemical fumigants (such as methyl bromide) cannot be used.
- ✰ Storage below 13°C will prevent insect feeding damage and reproduction. Storage at 5°C or below will control insect infestation.

18

Fig

Botanical Name: *Ficus carica*

Family: Moraceae

1. Introduction

Fig is one of the oldest cultivated fruits. It has a symbiotic relationship with insects for fruit setting. The fleshly fruit is consumed fresh or in dried form. It can also be canned or used for candy or jam making. It is delicious, wholesome, and nutrious fruit. Figs are a good source of carbohydrates, including fiber. The fruits are rich in calcium, iron, and vitamins A and C. Fresh or dried, they are valued for laxative properties. Medicinal uses such as applications against boils and other skin infections have been attributed to this fruit. Fig helps to maintain the acid-alkali balance of the body. In India its cultivation is confined to western parts of Maharashtra, Gujarat, Uttar Pradesh, Karnataka and Tamil Nadu.

2. Composition (Per 100g Edible Portion)

Constituent	Composition	
	Calimyrna	Mission
Protein (per cent)	3.00	2.82
Carbohydrates (g)	58.20	50.10
Fat (g)	1.90	0.90
Energy (cal)	253.00	212.00
Vitamin C (mg)	3.60	3.60
Vitamin B_1 (mg)	0.079	0.061
Vitamin B_2 (mg)	0.083	0.078
Vitamin A (IU)	142.00	92.00
Niacin (mg)	0.71	0.59
Calcium (mg)	174.00	130.00
Iron (mg)	2.50	2.40
Phosphorus (mg)	70.00	57.00
Magnesium (mg)	60.00	59.00
Copper (mg)	0.34	0.32
Zinc (mg)	0.48	0.43
Potassium (mg)	682.00	575.00
Sodium (mg)	10.00	10.00

3. Suitable Varieties

Poona, Conardia, Mission, Kadota, Brown, Turkey Calimyrna, Taranimt, Zidi, Black Mission

4. Maturity Indices

Fresh market figs must be harvested when almost fully ripe to be of good eating quality. Skin color and flesh firmness are dependable maturity and ripeness indices. 'Black Mission' figs should be light to dark purple rather than black and should yield to slight pressure. 'Calimyrna' figs should be yellowish-white to light yellow and firm.

5. Quality Indices

Fresh fig's skin color and flesh firmness are related to their quality and postharvest-life. Flavor is influenced by stage of ripeness and overripe figs can become undesirable due to fermentative products. Other quality

indices include absence of defects (such as bird-peck, sunburn, scab, skin break, and stem shrivel), insects, and decay.

6. Postharvest Management

Harvesting

Dried figs are harvested by picking them from the ground., where they fall after they have become ripe and somewhat dry. Mostly the picking is hand operated process. Fresh figs are hand picked from the tree, sorted carefully in small boxes and refrigerated for transport to retail markets.

Storage

(*a*) Optimum Temperature

-1°C to 0°C

Expedited forced-air cooling to 0°C is strongly recommended.

(*b*) Optimum Relative Humidity

90-95 per cent

(*c*) Rates of Respiration/Production of CO_2

Temperature °C	*ml CO_2/kg hr*
0	2–4
5	5–8
10	9–12
20	20–30

To calculate heat production, multiply ml CO_2/kg hr by 122 to get kcal/metric ton/day.

(*d*) Rates of Ethylene Production

Temperature	0°C	5°C	10°C	20°C
µl C_2H_4/kg hr	0.4–0.8	0.8–1.5	1.5–3.0	4.0–6.0

(*e*) Responses to Ethylene

Figs are slightly sensitive to ethylene action on stimulating softening and decay severity, especially if kept at 5°C or higher temperatures.

(*f*) Responses to Controlled Atmospheres (CA)

CA combinations of 5-10 per cent oxygen and 15-20 per cent carbon dioxide are effective in decay control, firmness retention, and reduction of respiration and ethylene production rates. Postharvest-life at optimum temperature and relative humidity depends upon cultivar and ripeness stage at harvest and ranges from 1 to 2 weeks.

7. Postharvest Disorders and Diseases

Physiological and Physical Disorders

CA-related Disorders

Extended storage in CA can result in loss of characteristic flavor. Figs exposed to less than 2 per cent oxygen and/or more than 25 per cent carbon dioxide develop off-flavors due to fermentative metabolism.

Pathological Disorders

Alternaria Rot

Caused by *Alternaria tenuis*, appears as small, round, brown- to-black spots over the fruit surface. Any cracks on the skin make the fruit more susceptible to the rot.

Black Mold Rot

Caused by *Aspergillus niger*, appears as dark or yellowish spots in the flesh with no external symptoms. At advanced stages the skin and flesh turn slightly pink color and white mycelia with black spore masses follow.

Endosepsis (Soft Rot)

Caused by *Fusarium moniliforms*, appears in the cavity of the fig making the pulp soft, watery and brown with sometimes an offensive odor.

Souring

Caused by various yeasts and bacteria, is a preharvest problem resulting from yeasts and bacteria carried into the figs by insects, especially vinegar flies, resulting in smells of alcohol or acetic acid.

Control of Postharvest Diseases

- Control of orchard insects to reduce fruit damage and transmission of fungi.
- Effective control of preharvest diseases.
- Strict sanitation of picking and transporting containers.
- Careful handling to minimize abrasions, cracks, and other physical damage. Do not pick figs for fresh market from the ground.
- Prompt cooling to 0°C and maintaining the cold chain all the way to the consumer.

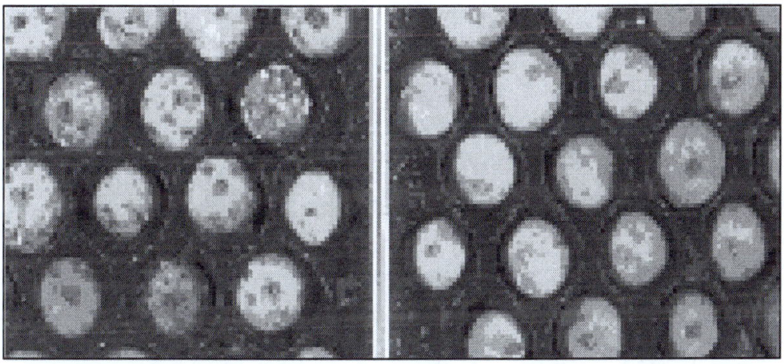

Air + 20 per cent CO_2 After 3 weeks 5°C

CO_2 Effect on Calimyrna

19

Jackfruit

Botanical Name: *Artocarpus heterophyllus*

Family: Moraceae

1. Introduction

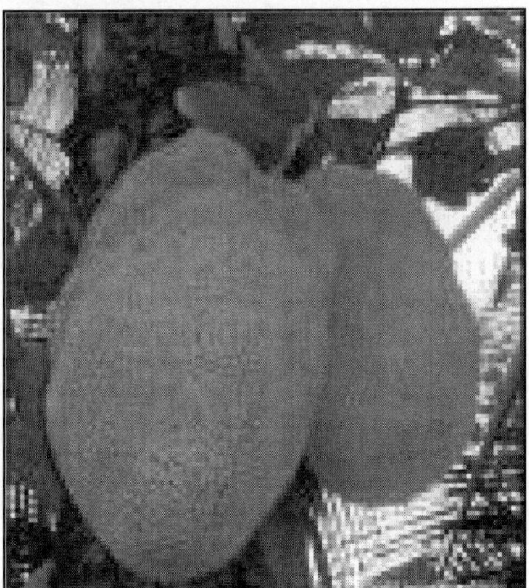

Jackfruit is popularly known as the poor man's food in the eastern and southern parts of India. A rich source of vitamin A, C, and minerals, it also supplies carbohydrates. Tender jackfruits are popularly used as vegetable. The skin of the fruit and its leaves are excellent cattle feed. Its timber is valued for furniture making since it is rarely attacked by white ants. The latex from the bark contains resin. Pickles and dehydrated leather are its preserved delicacies. Canning of flakes can be done. Nectar is prepared from its pulp. The rind is rich in pectin, which can be used for making jelly. The flakes, seeds sterile flowers, skin and core contain calcium pectate 4.6, 1.6, 3.7, 3.2 and 2.1 per cent respectively. They are considered as a good source of pectin.

2. Composition (Per 100g Edible Portion)

Constituent	Composition
Moisture (%)	76.2
Protein (g)	1.9
Carbohydrates (g)	19.8
Fat (g)	0.1
Fibre (g)	1.1
Minerals (g)	0.9
Energy (cal)	88
Vitamin c (mg)	7
Carotene (μg)	175
Thiamine (mg)	0.03
Riboflavin (mg)	0.13
Niacin (mg)	0.4
Calcium (mg)	20
Iron (mg)	0.56
Phosphorus (mg)	41

3. Suitable Varieties

Gulabi, Champa, hazar, NJT1, NJT2, NJT3, NJT4, NJC1, NJC2, NJC3 and NJC4.

4. Maturity Indices

Jackfruits can be in a very large size (as much as 90 cm long, 50 cm wide, and 25 kg in weight), depending on the cultivar, production area, and the fruit load on the tree. Color change from green to yellow to brown is used as an indication of maturity and ripeness stages. Optimum harvest for long-distance transport is when the fruit changes color from green to yellowish-green. Fruits are harvested with a portion of the stalk attached to be used in handling them.

5. Quality Indices

Fruit size, shape, color, and free from defects (sunburn, cracks, bruises) and decay.

Jackfruits contain 25-30 per cent carbohydrates (fresh weight basis) including about 15-20 per cent starch in unripe fruits that is converted to sugars (sucrose + glucose + fructose) in ripe fruits.

The unripe fruit is used as a starchy vegetable, either boiled or roasted. Average acidity is 0.25 per cent citric acid.

Fruitlet Quality

6. Postharvest Management

Harvesting

Tender fruits are harvested for use as vegetables until the seeds harden. Harvesting is done by cutting off the footstalks carrying the fruits.

7. Storage

(*a*) Optimum Temperature

13 ± 1°C; potential postharvest-life = 2-4 weeks, depending on cultivar and maturity stage.

(*b*) Optimum Relative Humidity

85-95 per cent

(*c*) Rates of Respiration/Production of CO_2

20-25 (preclimacteric) to 50-55 (climacteric peak) ml CO_2/kg·hr at 20°C

To calculate heat production multiplies ml CO_2/kg·hr by 122 to get kcal/metric ton/day.

(*e*) Responses to Ethylene

Exposure to 100ppm ethylene for 24 hours accelerates ripening of mature-green jackfruits at 20-25°C. During ripening, the starch is converted into sugars, the pulp color changes from pale white or light yellow to golden yellow, and the fruit aroma becomes intense.

8. Postharvest Diseases, Disorders and their Control

Physiological Disorders

Chilling Injury

Jackfruits exposed to temperatures below 12°C before transfer to higher temperatures exhibit chilling injury symptoms, including dark-brown discoloration of the skin, pulp browning and off flavor development, and increased susceptibility to decay.

20

Kiwifruit

Botanical Name: ***Actinidia deliciosa***

Family: Actinidiaceae

1. Introduction

Kiwin fruit is among the very few recent introductions, which have surpassed in popularity due to its tremendous commercial potential in the Sub-Himalayan region. A native to central China, it is being grown commercially in New Zealand, Italy, USA, China, Japan, Australia, France, Chile and Spain. In India, kiwi was first planted in the Lal Bagh Gardens at Bangalore as an ornmanental tree. With extensive research and development support, its commercial cultivation has been extended to the midhills of Himachal Pradesh, Uttar Pradesh, Jammu and Kashmir, Sikkim, Meghalaya, Arunachal Pradesh and Nilgiri hills.

2. Composition (Per 100g of edible Portion)

Constituent	Composition	Constituent	Composition
Energy (51 kcal)	217 kJ	Iron (µg)	800
Water (g)	83.8	Phosphorus (mg)	30
Protein (g)	1.0	Chloride (mg)	65
Lipid (g)	0.6	***Vitamins***	
Carbohydrate (g)	9.3	Carotene (µg)	370
Organic acids (g)	1.5	Vitamin B1(µg)	17
Fiber (g)	3.9	Vitamin B2(µg)	50
Minerals (g)	0.7	Nicotinamide (µg)	410
Minerals		Vitamin C (µg)	20-300
Sodium (mg)	1	***Carbohydrates***	
Potassium (mg)	295	Sucrose (mg)	1250
Magnesium (mg)	25	Glucose (mg)	4490
Calcium (mg)	40	Fructose (mg)	3540
		Organic acids	
		Malic acid (mg)	500
		Citric acid (mg)	990
		Oxalic acid	traces
		Salicylic acid (µg)	320

Ref: Deutsche Forschungsanstalt für Lebensmittelchemie, Garching bei München (ed), Der kleine "Souci-Fachmann-Kraut" Lebensmitteltabelle für die Praxis, WVG, Stuttgart 1991.

3. Suitable Varieties

Abott, Allison, Bruno, Hayward, Monty, Tomuri.

4. Maturity Indices

- ✰ Minimum of 6.5 per cent soluble solids content (SSC) at harvest.
- ✰ Minimum flesh firmness of 1kg (penetration force with an 8-mm). Late harvested kiwifruits retain their firmness better than early harvested fruit and have higher SSC at harvest and when ripe.

5. Quality Indices

- ✰ Freedom from growth cracks, insect injury, bruises, scars, sunscald, internal breakdown, and decay.

- ☆ Minimum of 14 per cent SSC when ripe (ready to eat); a kiwifruit at 2-3 lb flesh firmness is considered ripe.
- ☆ Kiwifruits are a good source of vitamin C.

6. Postharvest Management

Harvesting

The kiwi fruits are easily harvested by snapping off the fruit at the abscission layer at the base of the stalk. At least two pickings are made. Larger sized berries should be harvested first while smaller ones should be allowed to increase in size and improve the quality. After harvesting fruits are rubbed with coarse cloth to remove stiff hairs found on their surface.

Storage

(*a*) Optimum Temperature

0°C; highest freezing point is -1.5°C.

(*b*) Optimum Relative Humidity

90-95 per cent

(*c*) Rates of Respiration/Production of CO_2

Temperature	0°C	5°C	10°C	15°C	20°C
Rates of Respiration ml CO_2/kg hr	1.5-2.0	3-4	5-7	9-12	15-20

To calculate heat production multiply ml CO_2/kg hr by 122 to get kcal/metric ton/day.

(*d*) Rates of Ethylene Production

Less than 0.1 µl/kg·hr at 0°C, 0.1-0.5 µl/kg·hr at 20°C for unripe kiwifruit. Ripe kiwifruit (less than ½ kg firmness) produce 50-100 µl/kg·hr at 20°C.

(*e*) Responses to Ethylene

- ☆ Kiwifruits are extremely sensitive to ethylene. As little as 5-10 ppb ethylene will induce fruit softening.
- ☆ Avoid exposure of unripe kiwifruits to ethylene during harvest, transport, and storage.

(f) Responses to Controlled Atmospheres (CA)

Optimum CA 1-2 per cent O_2 + 3-5 per cent CO_2.

CA delays ripening and retains flesh firmness.

CO_2 levels above 7 per cent can cause internal breakdown of the flesh.

CA must be established within 2 days after harvest to maximize benefits; ethylene concentration should be kept below 20 ppb to avoid accelerated flesh softening and incidence of white core inclusions.

7. Postharvest Diseases, Disorders and their Control

Physiological Disorders

Freezing Damage

Flesh translucency starting at the stem end of the fruit and progressing toward the blossom end as the severity increases. Susceptible fruit become somewhat yellow fleshed with prolonged storage. Freezing damage can occur on early picked kiwifruit when stored at temperatures below 0°C.

Hard-Core

This disorder is induced by exposure of kiwifruit to ethylene plus carbon dioxide levels above 8 per cent. The fruit core fails to ripen when the remainder of the fruit is soft and ripe.

Internal Breakdown

These symptoms start as a slight discoloration (water soaking) at the blossom end of the fruit. With time this progresses around the blossom end and ultimately encompasses a large part of the fruit. As symptoms progress a "graininess" develops below the fruit surface beginning in the area around the blossom end of the fruit.

Pericarp Granulation

The occurrence of granulation is predominantly at the stylar end of the fruit, but as in the case of translucency may extend up the sides of fruit. This disorder also is more severe with prolonged storage and after ripening at 20°C. There is no obvious correlation between pericarp translucency and granulation since symptoms can occur independently.

Pericarp Translucency

This disorder has been noted in both air- and CA-stored kiwifruit at 0°C. It appears as translucent patches in the outer pericarp tissue at the stylar end which may extend up the sides of the fruit. Pericarp translucency is more severe after prolonged storage, but it can be observed after 12 weeks of storage at 0°C. The presence of ethylene in the storage atmosphere accelerates symptom development.

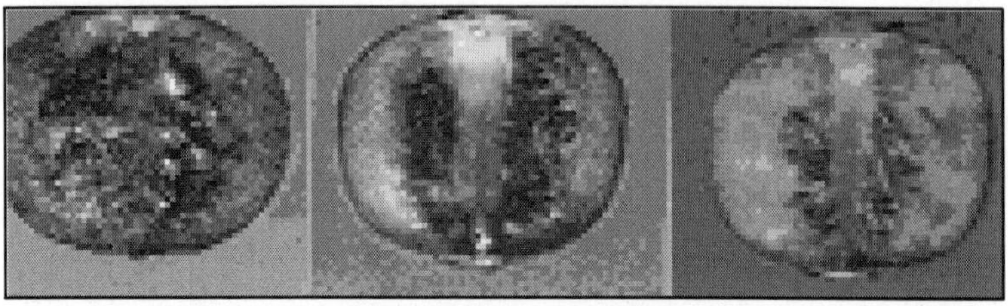

Starch Disappearance

Freezing Injury

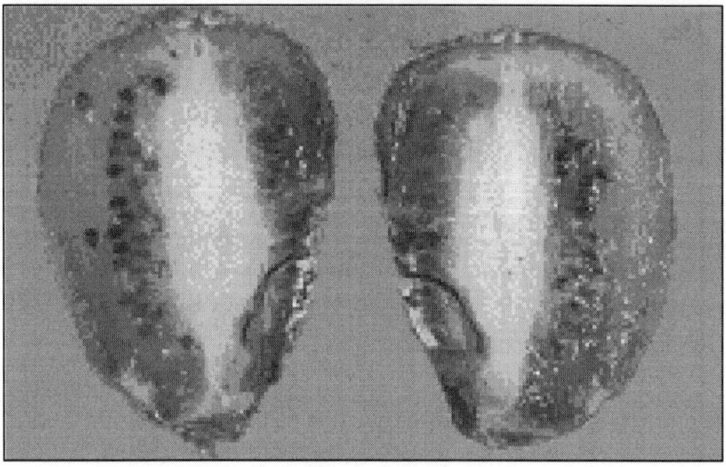

Alternaria Rot

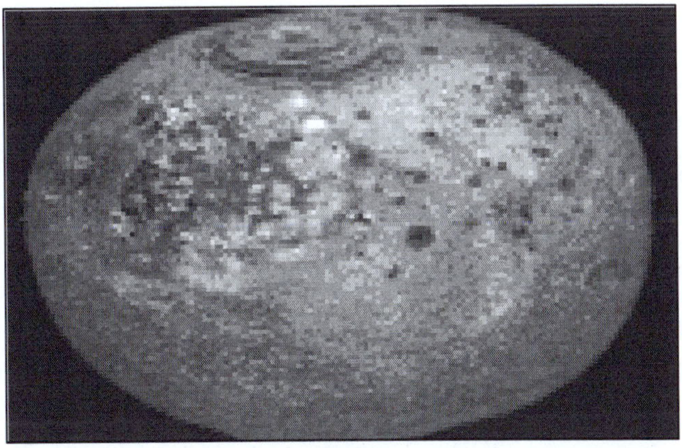

Botrytis Stem End Rot

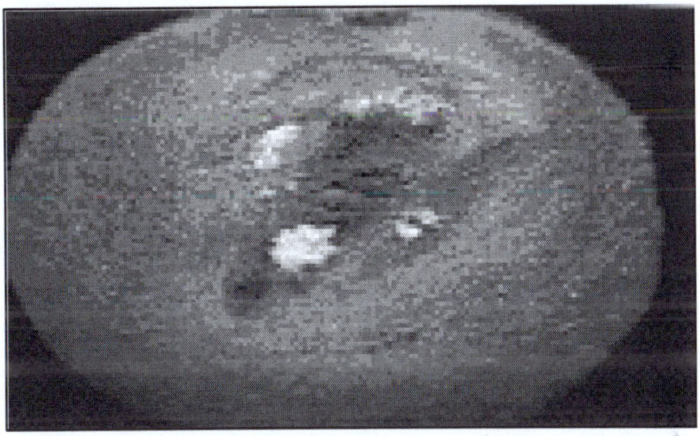

Penicillium Stem End

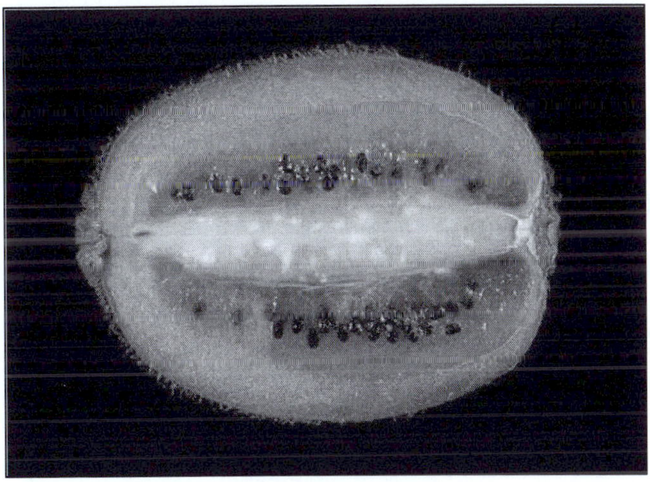

White Inclusion

White-Core Inclusions

The occurrence of white-core inclusions is directly related to the presence of ethylene in CA storage. This disorder results in distinct white patches of core tissue that are obvious in ripe fruit. Symptoms have been observed as early as 3 weeks after storage at 0°C.

Pathological Breakdown

Several pathogens can cause postharvest deterioration of kiwifruit. Botrytis gray mold rot caused by *Botrytis cinerea* is the most important and can directly invade the fruit or enter through wounds. Kiwifruit become much more susceptible to *Botrytis* (and other fungi) as they soften. Thus, maintaining fruit firmness (by rapid cooling, cold storage, and use of controlled atmospheres) can significantly reduce pathological breakdown. Sunburned fruit and physically damaged fruit are also more susceptible to postharvest diseases.

21

Litchi

Botanical Name: *Litchi chinensis*

Family: Sapindaceae

1. Introduction

Litchi was introduced to the tropical and subtropical world from China at the end of the seventeenth century and is now found situated within 15-20^0 latitude in most countries. Litchi is consumed fresh and in preserved forms. The edible portion of the fruit is a white-to cream-colored transulent pulp surrounding a glossy brown seed. The pulp is grapelike in texture, very succulent and aromatic, and it characterized by a sweet and acid taste. The most important production areas today are China, Taiwan, India, Pakistan, Thailand, Vietnam, Indonesia, Madagascar, South Africa, and Australia.

2. Composition (Per 100g Edible Portion)

Constituent	Composition
Moisture (per cent)	84.1
Crude fat (g)	0.2
Crude protein (g)	1.1
Carbohydrates (g)	13.6
Total sugars (g)	
Minerals (g)	0.5
Calcium (mg)	10
Phosphorus (mg)	35
Iron(mg)	0.7
Thiamine (mg)	0.02
Riboflavin (mg)	0.06
Niacin (mg)	0.4
Vitamin C (mg)	63
Crude fiber (g)	0.5
Energy K cal	61

3. Suitable Varieties

Early Seedless, Early Bedana, Rose Scented, Dehradun, Gulabi, Culcuttia, Shahi, Bombai, China, Late Bedana

4. Maturity Indices

Red Color (due to anthocyanins in the skin) is a good indicator of maturity along with fruit size (minimum of 25 mm in diameter).

Reaching the optimum range of sugar: acid ratio for the cultivar.

Litchi fruits should be harvested fully-ripe because they do not continue to ripen after harvest.

5. Quality Indices

Bright red color with no brown discoloration even though it is not an indicator of edible portion.

Quality Fruit

Sweet and juicy (edible portions); soluble solids: acid ratio of 30 or higher.

Free from defects (such as bird damage, insect damage, physical damage, cracking, and browning) and from decay.

Litchi fruits are excellent source of vitamin C (40 to 90 mg/100 g fresh weight)

6. Postharvest Management

Harvesting

Litchi fruits like other fruits are not harvested individually to avoid skin rupturing at the stem end quick rotting of fruits. They are harvested in bunches along with the portion of the branch and a few leaves.

Storagc

(*a*) Optimum Temperature

5°C, range: 1.5°C to 10°C depending on cultivar and intended storage duration

(*b*) Optimum Relative Humidity

90-95 per cent

Maintenance of high relative humidity is essential for reduction of water loss and browning, which is the major symptom of deterioration.

(c) Rates of Respiration/CO_2 Production

Temperature	5°C	10°C	20°C
ml CO_2/kg hr:	5-8	10-15	25-40

To calculate heat production multiply ml CO_2/kg hr by 122 to get kcal/metric ton/day.

(d) Rates of Ethylene Production

Less than 0.5 µl/kg·hr at 20°C

(e) Responses to Ethylene

Ethylene may accelerate deterioration of litchi fruits (breakdown of edible portion and increased decay).

(f) Special Treatment

Color can be preserved by bleaching with SO_2 fumigation followed by a dip in dilute hydrochloric acid.

(g) Responses to Controlled Atmospheres (CA)

Recommended CA: 3-5 per cent O_2 and 3-5 per cent CO_2.

Benefits include reduced skin browning and polyphenoloxidase activity and slower rates of losses of ascorbic acid, acidity, and soluble solids.

Modified atmosphere packaging is used commercially to a limited extent.

Exposure to oxygen levels below 1 per cent and/or carbon dioxide levels above 15 per cent may induce off-flavors and dull gray appearance of the pulp.

7. Postharvest Disorders, Diseases and their Control

Physiological and Physical Disorders

Pericarp Browning

Water loss (desiccation) of litchi fruits results in brown spots on the bright-red shell (pericarp). Under severe conditions or prolonged exposure, the spots enlarge and coalesce until the surface is completely brown. The flavor of the edible portion within browned fruit may or may not be adversely affected. Packaging in polymeric films reduces water loss and browning severity.

Chilling Injury

Symptoms include pericarp browning (similar to that caused by water loss) and increased susceptibility to decay. Storage at 1°C for 12 days before transfer to 20°C for one day results in pericarp browning.

Pericarp Split (Cracking)

Incidence and severity of cracking depend on cultivar and desiccation during storage. Cracks provide an entry way for decay-causing pathogens.

Edible Portion Breakdown

Prolonged storage and overmaturity may cause breakdown of edible portion (softening, loss of turgidity, translucency) and loss of flavor beginning at the blossom end and spreading to the stem end.

Pathological Disorders

Decay-inducing pathogens include *Alternaria* sp., *Aspergillus* sp., *Botryodiplodia* sp., *Colletotrichum* sp. and various yeasts.

Decay control can be achieved by reducing physical injuries to fruits and by prompt cooling and maintenance of the optimum temperature and relative humidity during litchi marketing.

Sun Burning

Fruit Cracking

Other decay control treatments under consideration include use of a 10-15 per cent CO_2-enriched atmosphere and biological control.

Disinfestation Treatments

Irradiation at 0.3 kGy can be used for insect disinfestation purposes with no adverse effects on litchi quality.

Exposure to heat at 45°C for 30 minutes can be used to control some insects on litchi fruits. Higher temperatures and/or longer exposures to heat damages the fruit.

Cold treatment (14 days at 1°C) may induce chilling injury (pericarp browning) in some cultivars.

22

Sapota or Sapodilla

Botanical Name: *Manikara achras*

Family: Sapotaceae

1. Introduction

Sapota or sapodilla is a native of tropical America, having originated in Mexico of Central America. It is a delicious fruit also known as Chickoo.

Fully ripe fruit is eaten as dessert. The pulp along with the skin can also be eaten, as the later is rich in nutritive value. A decoction is given in diarrhoea and in paludism. India is the largest producer of sapota. Although sapota is cultivated in India primarly for its edible fruit, it is cultivated in Mexico and Vannezuela mainly for the extraction of ckickle gum, resin derived from the bark. In India it has become popular fruit crop in Gujarat, Maharashtra, Karnataka, Tamil Nadu, Andhra Pradesh and Kerala.

2. Composition (Per 100g Edible Portion)

Constituent	*Composition*	
	Mature	*Ripe*
Starch (g)	0.90	0.71
Sucrose (g)	4.28	2.12
Glucose (g)	2.81	5.65
Fructose (g)	5.53	6.30
Crude Protein (T)	0.68	0.52
Protein (g)	0.51	0.50
Soluble Amino Acid (mg)	25	43
Total acidity (g)	0.15	0.11
Vitamin A (IU)	1603	410
Vitamin C (mg)	10	8
Dry matter (g)	21.9	23.4
Potassium (mg)	0.21	0.25
Phosphorus (mg)	8.0	11.6
Calcium (mg)	10.8	8.2
Iron (mg)	0.42	0.40

3. Suitable Cultivars

Cricket ball, Dwaraspudi, Kirtibarthi, Jonavalasa, Kalipati, Bhurupati, Pilipatti, Dhola Dhiwani, Jhumukhia, and Cricket Ball, Calcutta Special, Ever Bearer, Co1, Co2, PKM 1.

4. Maturity Indices

Skin color changes from light brown with a tinge of green to light brown to dark-brown.

Flesh color change from green to pinkish brown to reddish brown. (Can be examined through a small scratch on the surface).

5. Quality Indices

Appearance: size, shape, color, free from defects, and free from decay

Firmness (firm-ripe sapotes are preferred)

Flavor is related to soluble solids content (13-26 per cent) and acidity (0.2-0.3 per cent)

6. Postharvest Management

Harvesting

The sapota fruits are hand picked or harvested with a special harvester which has a round ring with a net bag fixed onto a long bamboo. Harvested fruits should be cleaned of latex and scurf by washing in clean water to make them look attractive. Fruits should be tightly packed in cardboard boxes of 10 kg capacity with rice straw as padding material and with ethylene absorbents and transported quickly to wholesale market.

Storage

(*a*) Optimum Temperature

14°C ± 1°C; storage potential is 2-4 weeks (depending on cultivar and and ripeness stage).

(*b*) Optimum Relative Humidity

90-95 per cent; packaging in perforated plastic bags or box liners reduces water loss at lower relative humidities.

(*c*) Rates of Respiration/CO_2 Production

Climacteric respiratory pattern; peak range = 25-35 ml CO_2/kg·hr at 20°C.

To calculate heat production multiplies ml CO_2/kg·hr by 122 to get kcal/metric ton/day.

(*d*) Rates of Ethylene Production

Peak range at 20°C = 2 4 µl/kg·hr

(*e*) Responses to Ethylene

Exposure of mature sapote fruits to 100 ppm ethylene for 24 hours at 20°C hastens their ripening. Removal of ethylene from the storage environment delays deterioration.

(*f*) Responses to Controlled Atmospheres (CA)

Storage in 5-10 per cent CO_2-enriched atmospheres delays ripening. Higher CO_2 concentrations may damage the appearance and taste of sapota.

7. Postharvest Disorders, Diseases and their Control

Physiological Disorders

Cold Storage Injury

Exposure to temperatures below 5°C for more than 10 days causes chilling injury as indicated by dark-brown spots on the peel, failure to ripen, off-flavor development, and increased decay incidence after transfer to higher temperatures.

Pathological Disorders

Anthracnose

Anthracnose caused by *Colletotrichum gloeosporioides* can be a serious problem in humid production areas. Effective preharvest control strategies reduce postharvest lesion development.

23

Strawberry

Botanical Name: *Fragaria ananassa*

Family: Rosaceae

1. Introduction

Strawberry is an attractive, luscious, tasty, and nutritious fruit, with a distinct and pleasant aroma, and delicate flavour. It has a unique place

among cultivated berry fruits. Rich in vitamin C and iron, it is mainly consumed as fresh. Jam and syrup is also prepared from strawberry. It is cultivated in tropical and sub tropical areas around the year. It is cultivated commercially in Himachal Pradesh, Uttar Pradesh, Maharashtra, West Bengal, Nilgiri Hills, Delhi, Haryana, Punjab, and Rajasthan. Owing to wide climate and soil adaptation and high returns, it has tremendous potential in India. Its cultivation can be to other suitable areas having assured irrigation and transport facilities.

2. Composition (Per 100g Edible Portion)

Constituent	Composition
Moisture (per cent)	87.8
Crude fat (per cent)	0.2
Crude protein (per cent)	0.7
Carbohydrates (per cent)	9.8
Total sugars (per cent)	
Minerals (per cent)	0.4
Calcium (mg/100g))	30
Phosphorus (mg/100g)	30
Iron(mg/100g)	1.8
Carotene (μg/100g)	18
Thiamine (mg/100g)	0.03
Riboflavin (mg/100g)	0.02
Niacin (mg/100g)	0.2
Vitamin C (mg/100g)	52
Crude fiber (per cent)	1.1
Energy K cal	44

3. Suitable Varieties

Chandlier, Tioga, Torrey, Selva, Belrubi, Fern, Oslograndi, Pajaro.

4. Maturity Indices

Based on berry surface color: Minimum 1/2 or 3/4 berry surface showing red or pink color, depending on grade.

5. Quality Indices

Appearance (color, size, shape, freedom from defects), firmness, flavor

Ripeness Stages

(soluble solids, titratable acidity and flavor volatiles), and nutritional value (Vitamin C). For acceptable flavor, a minimum of 7 per cent soluble solids and/or a maximum of 0.8 per cent titratable acidity are recommended.

6. Postharvest Management

Harvesting

Strawberries are harvested by hand picking or by using clippers. Strawberries are harvested in small trays or baskets. Berries should be picked along with small stem portion attached. Picking should be done in the morning; it will facilitate the better shelf life.

Precooling

After harvesting strawberries should be kept in shady place to avoid damage due to excessive heat in the open field. For distant marketing strawberries should be precooled at 4°C within 2 hours of harvesting and kept at the same temperature.

Storage

(*a*) Optimum Temperature

0 ± 0.5°C

(*b*) Optimum Relative Humidity

90 to 95 per cent

(*c*) Rates of Respiration/CO_2 Production

Temperature	0°C	10°C	20°C
ml CO_2/kg hr	6–10	25–50	50–100

To calculate heat production, multiply ml CO_2/kg hr by 122 to get kcal/metric ton/day.

(*d*) Rate of Ethylene Production

< 0.1 μl/C_2H_4/kg·hr at 20°C

(*e*) Response to Ethylene

Strawberries do not respond to ethylene by stimulation of ripening processes (strawberries should be harvested near to full ripe). Removal of ethylene from storage air may reduce disease development.

(*f*) Responses to Controlled/Modified Atmospheres

Modified atmosphere packaging for shipment with 10 to 15 per cent carbon dioxide reduces the growth of *Botrytis cinerea* (Grey Mold Rot) and reduces the respiration rate of the strawberries thereby extending postharvest life. Use of whole pallet covers for modified atmospheres is the most common method.

7. Postharvest Disorders, Diseases and their Control

Physiological Disorders

Perhaps because of rapid marketing and very short storage periods, physiological disorders are not a major concern with strawberry fruit.

Pathological Disorders

Diseases are the greatest cause of postharvest losses in strawberries. Postharvest fungicides are not used on strawberries; therefore, prompt cooling, storage at 0°C (32°F), preventing fruit injury, and shipment under

0°C 10°C 20°C

After 1 week

Temperature Effect

Abrasion Damage

Botrytis Rot

high carbon dioxide are the best methods for disease control. In addition, care should be taken to keep diseased or wounded berries out of trays at harvest as strawberry diseases will spread from diseased to nearby healthy berries (nesting).

Irradiation has been tested on strawberries for decay control with mixed success. Doses needed for adequate decay control without high carbon dioxide generally result in excessive berry softening.

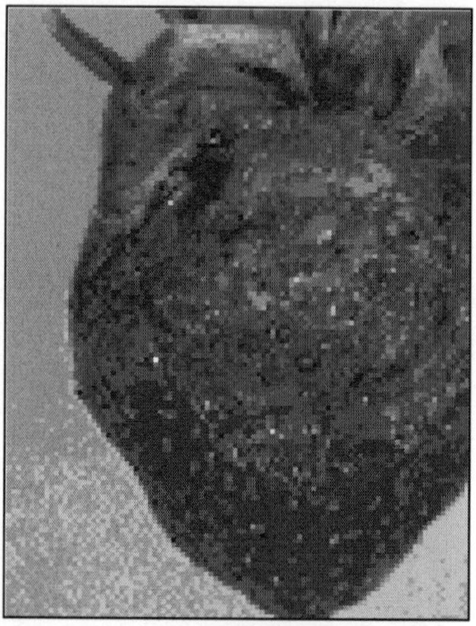

Anthracnose

Leather Rot

Mucor Rot

Botrytis Rot (Grey Mold)

Caused by *Botrytis cinerea* is the greatest cause of postharvest strawberry losses. This fungus continues to grow even at 0°C (32°), however growth is very slow at this temperature.

Rhizopus Rot

Caused by the fungus *Rhizopus stolonifer*. Spores of this fungus are usually present in the air and are easily spread. This fungus will not grow at temperatures below 5°C, therefore temperature management is the simplest method of control.

24

Lemon

Botanical Name: *Citrus limon*

Family: Rutaceae

1. Introduction

Lemons are commercially grown in tropical and sub tropical region of India. India ranks fifth among among lime and lemon producing countries in the world. It is cultivated almost in the all states of India. Andhra Pradesh, Maharashtra, Tamil Nadu, Karnataka, Gujarat, Bihar and Himachal Pradesh being major producing states. Lemons are less popular than limes in India. They are cultivated to a considerable extent commercially in Punjab, Rajasthan and tarai region of Uttar Pradesh.

2. Composition (Per 100g Edible Portion)

Constituent	Composition
Moisture (per cent)	85.0
Crude fat (g)	0.9
Crude protein (g)	1.0
Carbohydrates (g)	11.1
Total sugars (per cent)	
Minerals (g)	0.3
Calcium (mg)	70
Phosphorus (mg)	10
Iron(mg)	0.26
Thiamine (mg)	0.02
Riboflavin (mg)	0.01
Niacin (mg)	0.1
Vitamin C (mg)	39
Crude fiber (g)	1.7
Energy K cal	57

3. Suitable Varieties

Eureka, Lisbon, Villafrance, Lucknow Seedless, Kagzi Kalan, Nepali Oblong, Nepali Round, Pant Lemon 1.

4. Maturity Indices

A minimum juice content by volume of 28 or 30 per cent depending on grade; color lemons picked at the dark-green stage have the longest postharvest life while those picked fully-yellow must be marketed more rapidly.

5. Quality Indices

Yellow color intensity and uniformity; size; shape; smoothness; firmness; free should be from decay; and freedom from defects including freezing damage, drying, mechanical damage, rind stains, red blotch, shriveling, and discoloration.

6. Postharvest Management

Harvesting

The lemon fruits should be neither plucked nor torn off with clippers.

Generally limes and lemons are harvested with a pole harvester, having an iron hook and a net on one end.

Storage

(*a*) Optimum Temperature

12-14°C depending on cultivar, maturity-ripeness stage at harvest, production area, and duration of storage and transport (can be up to 6 months).

(*b*) Optimum Relative Humidity

90-95 per cent

(*c*) Rates of Respiration/CO_2 Production

Temperature	10°C	15°	20°
ml CO_2/kg hr	5-6	7-12	10-14

To calculate heat production multiply ml CO_2/kg hr by 440 to get Btu/ton/day or by 122 to get kcal/metric ton/day.

(*d*) Rates of Ethylene Production

< 0.1 µl/kg·hr at 20°C

(*e*) Responses to Ethylene

If degreeing is desired, lemons can be treated with 1-10 ppm ethylene for 1-3 days at 20 to 25°C, but this exposure may accelerate deterioration rate and decay incidence

(*f*) Responses to Controlled Atmospheres (CA)

CA of 5-10 per cent O_2 and 0-10 per cent CO_2 can delay senescence including loss of green color of lemons. Fungistatic CO_2 levels (10-15 per cent) are not used because they may induce off-flavors due to accumulation of fermentative volatiles, especially if O_2 levels are below 5 per cent. Removal of ethylene from lemon storage facilities can reduce rate of senescence and decay incidence.

7. Postharvest Disorders, Diseases and their Control

Physiological Disorders

Chilling Injury

Symptoms include pitting, membranous staining, and red blotch. Severity depends upon cultivar, production area, harvest time, maturity-ripeness stage at harvest, and time-temperature of postharvest handling operations. Moderate to severe chilling injury is usually followed by decay.

Oil Spotting (Oleocellosis)

Breaking of oil cells due to physical stress on turgid fruits causes release of the oil that damages surrounding tissues. Avoiding harvesting lemons when they are very turgid and careful handling reduces severity of this disorder.

Pathological Disorders

Green Mould

Caused by *Penicillium digitatum*, which penetrates the fruit rind through wounds. Symptoms begin as water-soaked area at the fruit surface followed by growth of colorless mycelium, then sporulation (green color).

Blue Mould

Caused by *Penicillium italicum*, which can penetrate the uninjured peel and can spread from one lemon to adjacent lemons. Symptoms are similar to green mold except that the spores are blue.

Altenaria Rot

Caused by *Alternaria citri*, which enters the lemons through their buttons. Preharvest treatment with gibberellic acid or postharvest treatment with 2,4-D delay senescence of the buttons and subsequent decay by Alternaria.

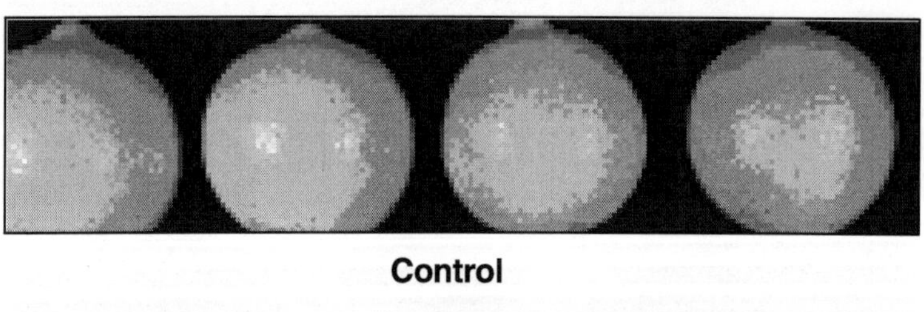

Control

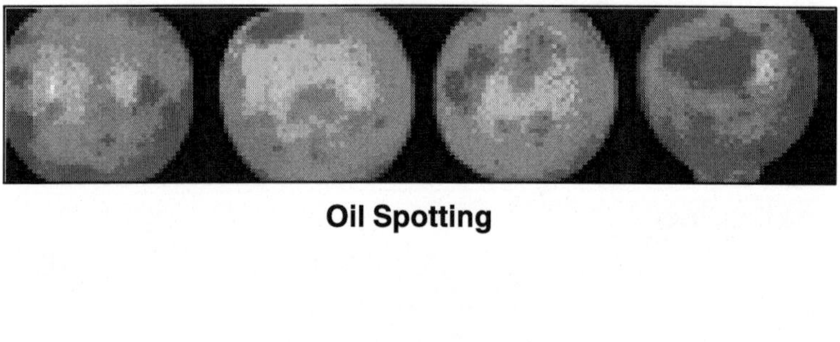

Oil Spotting

Control Strategies

- ☆ Careful handling during harvesting and handling to minimize cuts, scratches, and bruises.
- ☆ Treatment with postharvest fungicides and/or biological agents.
- ☆ Prompt cooling to the proper temperature range.
- ☆ Maintaining optimum ranges of temperature and relative humidity and exclusion of ethylene during transport and storage.
- ☆ Effective sanitation throughout the handling system.

25

Watermelon

Botanical Name: ***Citrullus lanatus***

Family: Cucurbitaceae

1. Introduction

Watermelon is an important cucurbitaceous vegetable. It is known as tarbuj, tarmuj, kalinda and kalindi in different parts of India. Though it can be grown in garden land, it is a major river-bed crop of Uttar Pradesh, Rajasthan, Gujarat, Maharashtra and Andhra Pradesh. As a common summer season crop, it is grown from the lower Himalayan region to southern parts of India, Punjab, Haryana, Karnataka, Assam, West Bengal, Orissa, Himachal Pradesh, Uttar Pradesh, Tamil Nadu and Rajasthan being major watermelon growing states. An excellent refreshing and cooling beverage after adding a pinch of salt and black pepper.

2. Composition (Per 100g of Edible Portion)

Constituent	Composition
Moisture (%)	95.8
Protein (g)	0.2
Fat g(g)	0.2
Minerals (g)	0.2
Fiber (g)	0.2
Carbohydrate (g)	3.3
Energy (K cal)	16
Calcium (mg)	11
Phosphrous (mg)	12
Iron (mg)	12
Carotene (mg)	7.9
Thiamine (mg)	0.02
Riboflavin (mg)	0.04
Niacin (mg)	0.1
Vitamin C (mg)	1

3. Suitable Cultivars

Arka Jyoti, Arka Manik, Asahi Yamato, Durgapura Kesar, Durgapura Meetha, Improved Shipper, New Hampshire Midget, Pusa Bedana, Special No. 1, Sugar Baby.

4. Maturity Indices

Watermelons are harvested at full maturity as they typically do not develop in internal color or increase in sugars after being removed from the vine. The ground spot (the portion of the melon resting on the soil) changes from pale white to a creamy yellow at proper harvest maturity. Another indicator used at harvest include a wilted but not fully desiccated vine tendril proximal to the stem-end attachment. Destructive sampling is used to judge maturity of a population of watermelons. For seeded cultivars, maturity is reached when the gelatinous covering (aril) around the seed is no longer apparent and the seed coat is hard. Cultivars vary widely in soluble solids at maturity. In general, a soluble solids content of at least 10 per cent in the flesh near the center of the melon is an indicator of proper maturity if the flesh is also firm, crisp and of good color.

5. Quality Indices

Watermelons should be symmetrical and uniform in appearance. The surface should be waxy and bright in appearance. Absence of scars, sunburn, transit abrasions or other surface defects or dirt. No evidence of bruising. Appears heavy for size.

6. Postharvest Management

Harvesting

Mostly the fruits become fit for harvesting 30-40 days after anthesis. These should be separated from the vine with the help of knife.

Storage

(*a*) Optimum Temperature

10–15°C. Storage life is typically 14 days at 15°C with up to 21 days attainable at 7-10°C. For short-term storage or transit to distant markets (> 7 days), most recommendations use 7.2°C and 85-90 per cent R.H. as the acceptable handling conditions. Watermelons are, however, prone to chilling injury at this temperature. Extended holding at this temperature will induce chilling injury, rapidly evident after transfer to typical retail display temperatures. Due to their high volume many watermelons are still shipped without precooling or refrigeration during transit. These fruit must be utilized for prompt market sales as quality declines rapidly under these conditions.

(*b*) Optimum Relative Humidity

85-90 per cent; High relative humidity is generally advisable to reduce desiccation and loss of glossiness.

(*c*) Rates of Respiration/CO_2 Production

Temperature °C	*ml CO_2/kg hr*
0	NR
5	3–4
10	6–9
15	NA
20	17–25
25	NA

To calculate heat production multiply ml CO_2/kg hr by 122 to get kcal/metric ton/day.

NR: Not recommended due to chilling injury; NA: Not available.

(*d*) Rates of Ethylene Production

Low–0.1–1.0 µl/kg·hr at 20°C

(*e*) Responses to Ethylene

Exposure to an ethylene concentrations as low as 5ppm for 7 days at 18°C will cause unacceptable loss of firmness and eating quality.

(*f*) Responses to Controlled Atmospheres (CA)

Controlled atmosphere storage or shipping is not recognized as offering. Controlled benefits for watermelon.

7. Postharvest Disorders, Diseases and their Control

Physiological Disorders

Chilling Injury

Typically occurs after storage at temperatures < 7°C. Symptoms of chilling injury include pitting, decline in flesh color, loss of flavor, off-flavors and increased decay when returned to room temperatures.

Physical Injury

Improper handling and loading of bulk watermelons too often result

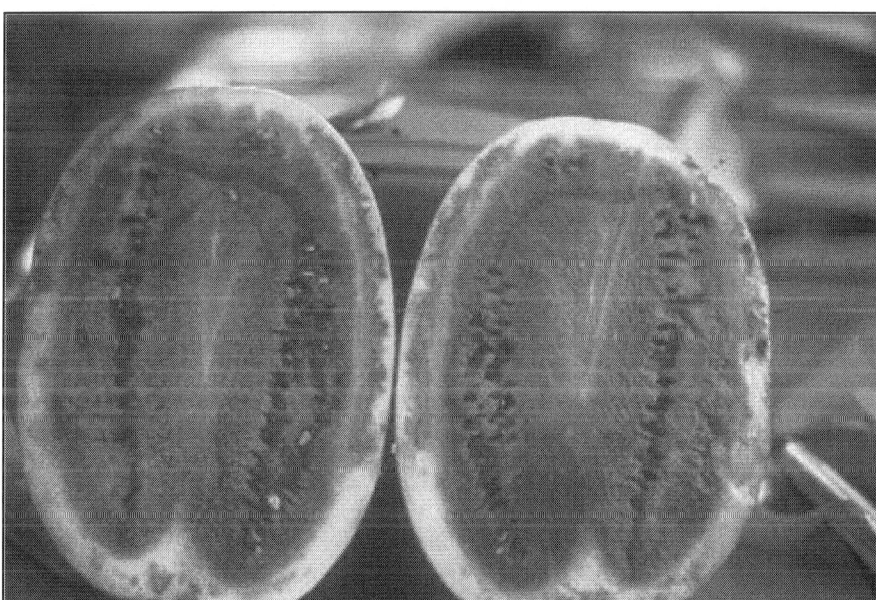

Anthracnose

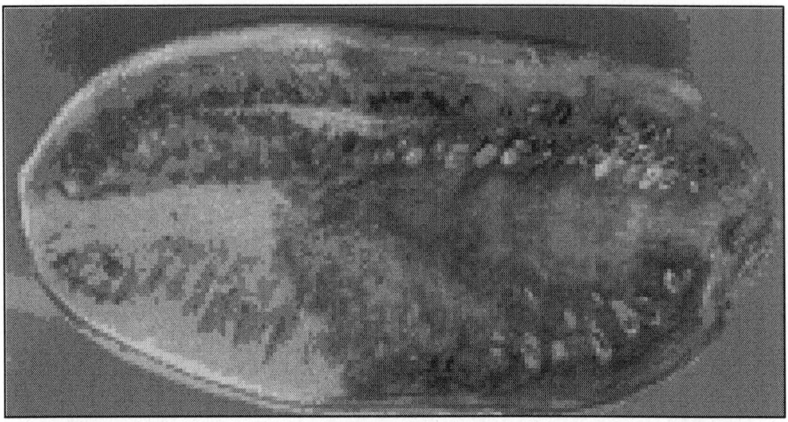

Lasiodiplodia Rot

Botrytis Stem End Rot

in serious transit losses due to bruising and cracking. Internal bruising leads to premature flesh breakdown and mealiness.

Pathological Disorders

Disease can be an important source of postharvest loss depending on season, Disorders region and local climatic conditions at harvest. Generally these losses are low in comparison with physical injury due to bruising and rough handling. Black Rot, caused by *Didymella bryoniae,* Anthracnose caused by *Colletotrichum orbiculare,* and *Phytophthora* Fruit Rot are common in areas with high rainfall and humidity during production and harvest. An extensive list of stem-end, blossom-end, rind decay or surface lesions may occur, including the bacterium *Erwinia* and the fungal pathogens *Alternaria, Botrytis, Cladosporium, Geotrichum, Rhizopus,* and occasionally *Mucor, Fusarium,* and *Tricothecium.*

Important to Note

Cut watermelon for slices or cubes for fresh-cut fruit salads have a very short period of optimal quality. Flesh becomes water-soaked and mealy.

26
Musk Melon

Botanical Name: ***Cucumis melo*** **L.**

Family: Cucurbitaceae

1. Introduction

Muskmelons are primarily used as fresh fruit. The unripe fruit may be cooked as vegetable by the village folk. Rich in vitamin A, B and C, and calcium, phosphorus and iron, muskmelons provides a wholesome food. Seed kernels are edible, tasty and nutritious, since they are rich in oils and energy.

2. Composition (Per 100g of Edible Portion)

Constituent	*Composition*
Moisture (g)	95.2
Protein (g)	0.3
Fat (g)	0.2
Minerals (g)	0.4
Fiber (g)	0.4
Carbohydrate (g)	3.5
Energy (K cal)	17
Calcium (mg)	32
Phosphrous (mg)	14
Iron (mg)	1.4
Carotene (mg)	169
Thiamine (mg	0.11
Rlbotlavin (mg)	0.08
Niacin (mg)	0.3
Vitamin C (mg)	26

3. Suitable Varieties

Arka Jeet, Arka Rajhanse, Durgapura Madhu, Hara Madhu, Hissar Madhur, Hisar Saras, MH 10, Punjab Hybrid, Pusa Rasraj, Punjab Rasila, Pusa Sharbati, Punjab Sunheri.

4. Maturity Indices

Muskmelons are harvested by maturity and not by size. Commercial maturity is ideally at the firm-ripe stage or "3/4 to full-slip" when a clear abscission (slip, separation) from the vine occurs with light pressure. Muskmelons ripen after harvest but do not increase in sugar content.

Cultivars vary in their external color at this stage of maturity and may retain a greenish cast. This skin color typically transitions from gray to dull green when immature, deep uniform green at maturity, and light yellow at full ripeness. A raised and well-rounded netting on the fruit surface is another indicator of proper commercial maturity.

5. Quality Indices

Well-shaped nearly spherical and uniform in appearance. Smooth stem end with no adhering peduncle (stem-attachment), which suggests premature harvest. Absence of scars, sunburn or surface defects. Firm with no evidence of bruising or excessive scuffing. Appears heavy for size and has firm internal cavity without loose seeds or liquid accumulation.

Distinction among grades is based predominantly on external appearances and measured soluble solids, minimum of 11 per cent soluble solids.

6. Postharvest Management

Harvesting

Fruits maturing on the vine, without becoming overripe are superior in quality to those harvested immature, or left on the vine after they have become mature.These should be separated from the vine with the help of knife.

Precooling

Rapid precooling soon after harvest is essential for optimal postharvest keeping quality. The precooling endpoint is typically 10°C but 4°C is more desirable. Forced-air cooling is the most common practice but hydrocooling is also utilized.

Storage

(*a*) Optimum Temperature

2.2°–5°C

Storage life is up to 21 days at 2.2°C but sensory quality may be reduced. Typically 12-15 days of shelf life are attainable within the optimum range. Short term storage or transit temperatures below this range are used by some in the trade but may result in chilling injury after several days [for example, 7 days or longer at temperatures below 2.2°C].

(*b*) Optimum Relative Humidity

90–95 per cent; High relative humidity is essential to maximize postharvest quality and prevent desiccation. Water loss through scuffed and damaged surface netting can be significant. Extended periods of higher humidity or condensation may encourage the growth of stem-scar and surface molds.

(*c*) Rates of Respiration/CO_2 Production

Temperature	0°C	5°C	10°C	15°C	20°C	25°C
ml CO_2/kg hr	2–3[NR]	4–5	7–8	17–20	23–33	65–71

To calculate heat production multiply ml CO_2/kg h by 122 to get kcal/metric ton/day.

NR: Not recommended for more than a few days due to chilling injury.

(*d*) Rates of Ethylene Production

Intact fruit–40–80µl/kg·h at 20°C

Production Fresh-cut–7-10µl/kg·h at 5°C

(*e*) Responses to Ethylene

Muskmelons are moderately sensitive to exogenous ethylene and over-ripening may be a problem during distribution and short-term storage.

(*f*) Responses to Controlled Atmospheres (CA)

Controlled atmosphere storage or shipping offer only moderate benefits for muskmelons under most conditions. With extended transit times (14-21 Atmospheres (CA) days), muskmelons are reported to benefit from delayed ripening, reduced respiration and associated sugar loss, and inhibition of surface molds and decay. Consensus atmospheres of 3 per cent O_2 and 10 per cent CO_2 at 3°C has been demonstrated. Elevated CO_2 at 10-20 per cent is tolerated but will cause effervescence in the fruit flesh. This carbonated flavor is lost on transfer to air.

Low O_2 (<1 per cent) or high CO_2 (> 20 per cent) will cause impaired ripening, off-flavors and odors, and other condition defects.

7. Postharvest Diseases, Disorders and their Control

Physiological Disorders

Chilling injury typically occurs after storage at temperatures < 2°C Disorders for several days. Sensitivity to chilling injury decreases as melon maturity and ripeness increases. Symptoms of chilling injury include pitting or sunken areas, failure to ripen, off-flavors and increased surface decay.

Pathological Disorders

Disease can be an important source of postharvest loss depending on season, region and handling practices. Commonly, decay or surface lesions result from the fungal pathogens *Alternaria, Penicillium, Cladosporium,*

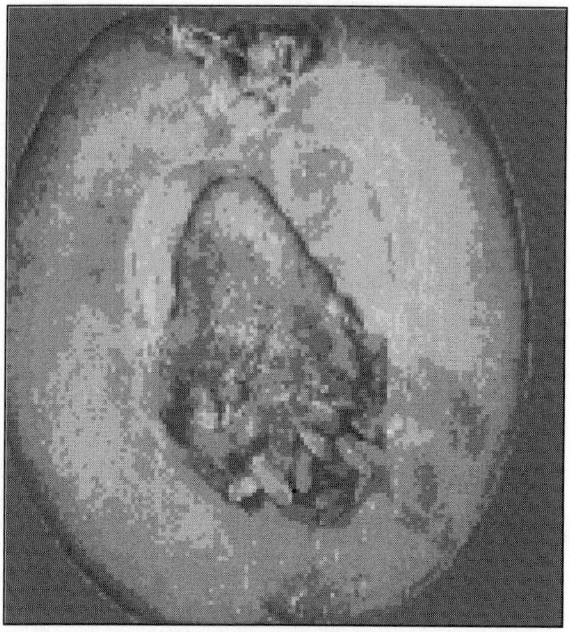

Sour Rot

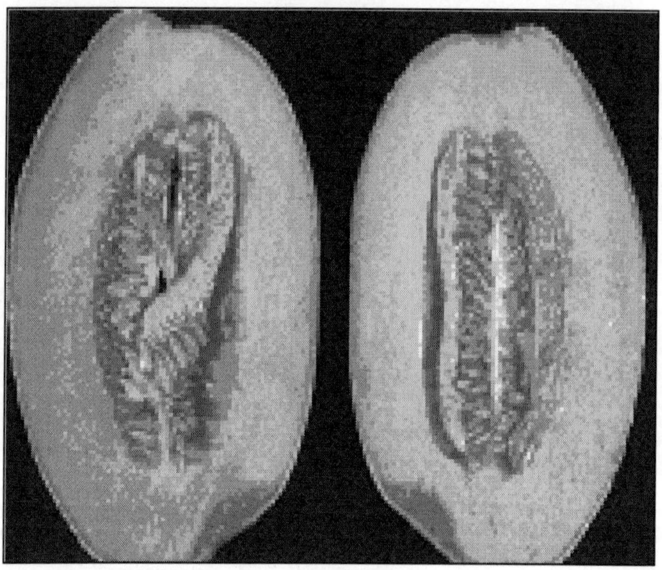

Rhizopus Rot

Geotrichum, Rhizopus, and to a lesser extent *Mucor*. Treatment with hot air or hot water immersion (55°C for 0.5–1.0 min.) has been effective in preventing surface mold but has not been used extensively for commercial treatments. CA can be effective in delaying fungal growth on the stem-end and fruit surface.

Vegetables

27

Capsicum

Botanical Name: *Capsicum annuum*

Family: Solanaceae

1. Introduction

Capsicum also known as well as bell pepper or sweet pepper, is a popular vegetable in India. It is mainly cultivated in Himachal Pradesh, Uttar Pradesh, Gujarat, Maharashtra, Karnataka, Ranchi region of Bihar, and hilly regions of Tamil Nadu. It grows well in summer season in hills and cooler season in the plains.

2. Composition (Per 100 gm of Edible Portion)

Constituent	*Composition*
Water (%)	92.4
Food value calorie	29
Protein (gm)	1.2

Constituent	Composition
Calcium (mg)	11
B1 and B20 (gm)	02-0.1
Folic acid (gm)	1.32-2.9
Nicotinic acid(gm)	6-10
Vitamin A (IU)	870
Ascorbic acid (gm)	175
Thiamine (gm)	0.06
Riboflavin (gm)	0.03
Niacin (gm)	0.55
Vitamin C (gm)	3.21

3. Suitable Varieties

Arka Mohina, Arka Gaurav, Arka Basant, California wonder, Pusa Deepti Chinese giant, world Beater, Yolo wonder, Hungrein wax. Bharat, Indra, Sun 1090 and Green Gold are popular hybrids. Open pollinated capsicums still dominate the market but more recently a number of hybrids are becoming popular among the growers. Hybrids are superior in yield popular fruits.

4. Maturity Indices

Green Peppers: fruit size, firmness, color

Colored Peppers: minimum 50 per cent coloration

5. Quality Indices

- ☆ Uniform shape, size and color typical of variety
- ☆ Firmness
- ☆ Should be free from defects such as cracks, decay, sunburn

6. Postharvest Management

Harvesting

Sweet Pepper are picked with an upward twist, which leaves a piece of stem attached. Young immature peppers are rather soft and yield readily to mild pressure of the finger. Green fruits ready for harvest are relatively firm and crisp.

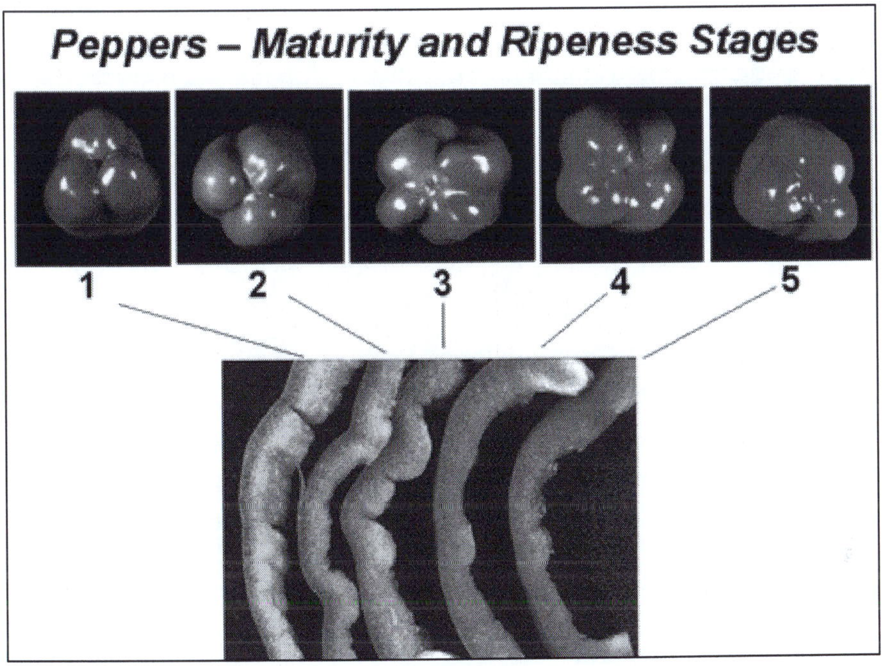

Maturity

Modified chippers developed at CIPHET, Abohar can be used for efficient and safe harvesting of capsicum.

Storage

(*a*) Optimum Temperature

Peppers should be cooled as soon as possible to reduce water loss. Peppers stored above 7.5°C suffer more water loss and shrivel. Storage at 7.5°C is best for maximum shelf-life (3-5 weeks); peppers can be stored at 5°C for 2 weeks, and although this reduces water loss, chilling injury will begin to appear after that period. Symptoms of chilling injury include pitting, decay, discoloration of the seed cavity, softening without water loss. Ripe or colored peppers are less chilling sensitive than green peppers.

(*b*) Optimum Relative Humidity

> 95 per cent; firmness of peppers is directly related to water loss.

(*c*) Rates of Respiration/Co_2 Production

Temperature	5°C	10°C	20°C
ml CO_2/kg hr	3–4	5–8	18–20

To calculate heat production, multiply ml CO_2/kg hr by 122 to get kcal/metric ton/day.

(*d*) Rates of Ethylene Production

Bell peppers are nonclimacteric in behaviour and produce very low levels of ethylene: 0.1-0.2 µl/kg·hr at 10°-20°C.

(*e*) Responses to Ethylene

Bell Peppers respond very little to ethylene; to accelerate ripening or color change, holding partially colored peppers at warm temperatures of 20-25°C with high humidity (>95 per cent) is most effective.

(*f*) Responses to Controlled Atmospheres (CA)

Peppers generally do not respond well to CA. Low O_2 atmospheres (2-5 per cent O_2) alone have little effect on quality and high CO_2 atmospheres (>5 per cent) can damage peppers (pitting, discoloration, softening) especially if they are stored below 10°C. Atmospheres of 3 per cent O_2+ 5 per cent CO_2were more beneficial for red than green peppers stored at 5°C to 10°C for 3-4 weeks.

(*g*) Shrink Wrapping and Storage Life

Results of various experiments conducted at CIPHET, Abohar reveals that film wrapping of capsicum could be an alternative for controlling water loss, spread of decay and retantion of bell shape without adverse effect on flavour and colour development. Tray or individual cover wrapping with 15µ heat shrinkable film and storage at cold storage was most effective in reducing the weight loss, decay and maintain the quality in terms of TSS, acidity, vitamin C, colour aroma and flavour.

7. Postharvest Disorders, Diseases and their Control

Physiological Disorders

Blossom End Rot

This disorder occurs as a slight discoloration or a severe dark sunken lesion at the blossom end; it is caused by temporary insufficiencies of water and calcium and may occur under high temperature conditions when the peppers are rapidly growing.

Pepper Speck

This disorder appears as spot-like lesions that penetrate the fruit wall; cause is unknown; some varieties are more susceptible than others.

Cold Storge Chilling Injury

Symptoms include surface pitting, water-soaked areas, decay (especially *Alternaria*), and discoloration of the seed cavity.

Pathological Disorders

The most common decay organisms are *Botrytis, Alternaria,* and soft rots of fungal and bacterial origin.

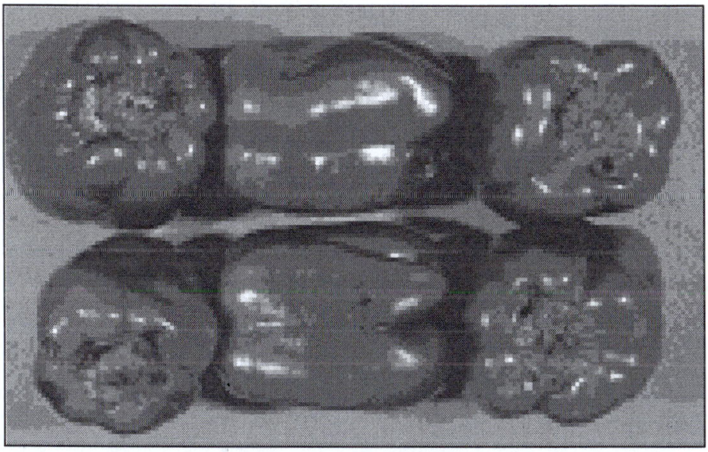

Chilling Injury Symptoms

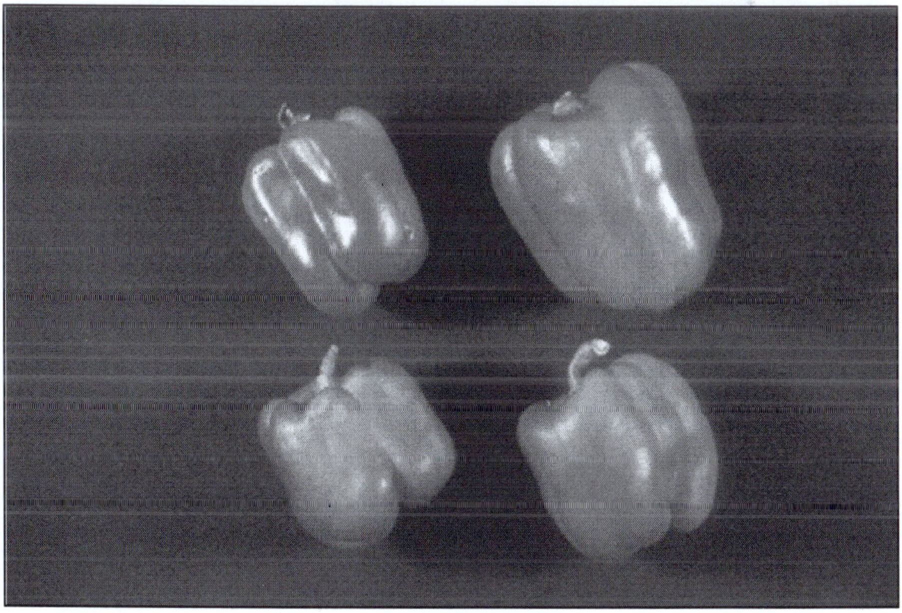

Solar Yellowing

Botrytis or Grey Mold Decay

This is a common decay-causing organism on peppers; field sanitation and prevention of wounds on the fruit helps to reduce its incidence. Botrytis will grow well at the recommended storage temperatures. High CO_2levels (>10 per cent) which can control *Botrytis* damage peppers. Hot water dips of peppers can effectively control botrytis rot (55°C water for 4 minutes) without causing fruit injury.

Alternaria Rot

The presence of black Alternaria rot, especially on the stem end of the pepper is a symptom of chilling injury; best control measure is to store at 7.2°C.

Bacterial Soft Rot

Soft rotting areas can be caused by several bacteria which attack damaged tissue; soft rots can also be common on washed or hydrocooled peppers where water sanitation was deficient.

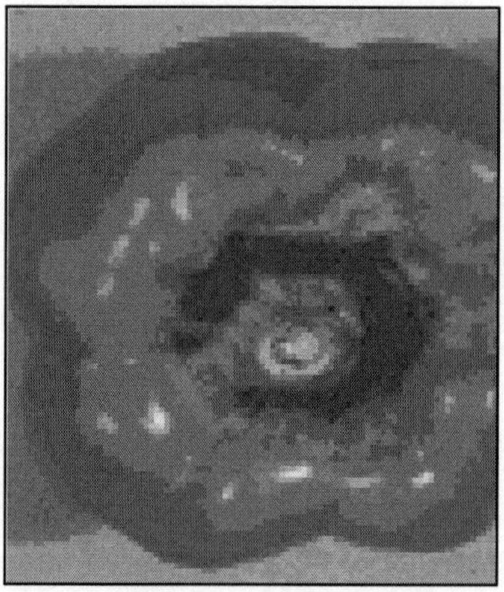

Alternaria Rot

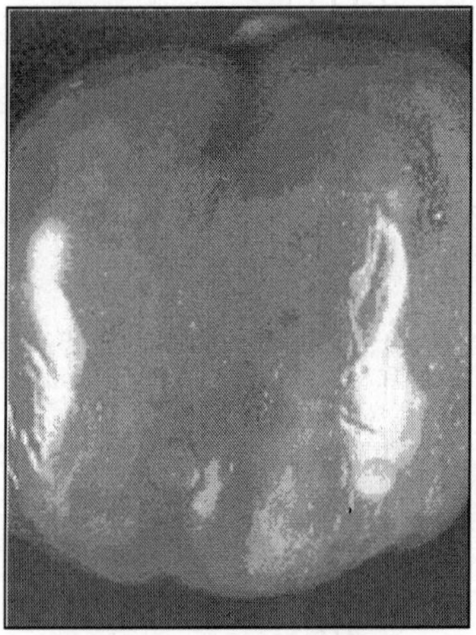

Bacterial Soft Rot

Other Common Postharvest Defects

Mechanical damage (crushing, stem punctures, cracks, etc.) is very common on peppers; physical injury not only detracts from the visual quality of the peppers but also causes increased weight loss and decay.

28
Cabbage

Botanical Name: ***Brassica oleracea*** **var** ***capitata***

Family: Cruciferaeae

1. Introduction

Cabbage is another important vegetable of cole group. A rich source of Vitamin A, B and C, it also contains minerals. It covers about 4 per cent of total area under vegetables. India comes next to China in cabbage production. It is now grown almost throughout the year. Orissa, West Bengal, Bihar, Karnataka, Maharashtra, Gujarat and Punjab are major cabbage-growing states. Commonly grown cabbage in India is white. The heads of cabbage vary from flat-topped to long –oval. Varieties with compact, round heads are preferred though pointed-head varieties are also grown. The tender leaves are primarily used as cooked vegetable, more in raw than in processed form.

2. Composition (Per 100g Edible Portion)

Constituent	Composition
Moisture (%)	91.1
Protein (g)	1.8
Fat (g)	0.1
Minerals (g)	0.6
Fiber (g/100g)	1.0
Carbohydrate (g)	4.6
Energy (K cal)	27
Calcium (mg)	39
Phosphrous (mg)	44
Iron (mg)	0.8
Carotene (mg)	120
Thiamine (mg)	0.06
Riboflavin (mg)	0.09
Niacin (mg)	0.4
Vitamin C (mg)	124
Folic acid (mg)	33

3. Suitable Varieties

Copenhagen market, Drumhead Savoy, Golden Acre, Pride of India, Pusa Drumhead, Pusa Mukta, Red cabbage.

Round hard cabbages and Chinese (also called Napa) cabbages are from the same genus (*Brassica*) but different species (*B. oleracea* var *capitata* = cabbage, *B. campestris* var. *pekinensis* = Chinese cabbage). Chinese cabbages may be cylindrical or rounded and may be less compact than round cabbages.

4. Maturity Indices

Maturity is based on head compactness. A compact head can be only slightly compressed with moderate hand pressure. A very loose head is immature, and a very firm or hard head is mature.

5. Quality Indices

After trimming outer wrapper leaves, cabbage heads should be a color typical of the cultivar (green, red, or pale yellow-green), firm, heavy for the size and free of insect, decay, seed stalk development and other defects. Leaves should be crisp and turgid.

6. Postharvest Management

Harvesting

The cabbage heads are harvested with the help of particularly designed knife. The heads should be harvested when they attain full size, after that they burst or lose.

Storage

Before storage cabbage heads should be trimmed and washed. For washing, machines developed at PAU, Ludhiana given better results.

(*a*) Optimum Temperature and Relative Humidity

Most cabbage is room cooled. Storage at 0°C with >95 per cent RH is required to optimize cabbage storage life. Early crop round cabbage can be stored 3-6 weeks, while late crop cultivars can be stored for up to 6 months. For the latter, storage at -0.5°C is sometimes recommended. Chinese cabbage can be stored from 2 to 6 months, depending on cultivar, at 0° to 2.5°C. Deterioration of cabbage during storage is associated with stem or seed stalk growth (bolting), root growth, internal breakdown, leaf abscission, discoloration, decay and black speck. Long-term storage usually results in extensive trimming of heads to remove deteriorated leaves.

(*b*) Freezing Injury

Freeze damage appears as darkened translucent or water-soaked areas that will deteriorate rapidly after thawing. Freeze damage can occur if round cabbages are stored below -0.9°C.

(*c*) Rates of Respiration/CO_2 Production

Round and Chinese cabbages have similar moderately low respiration rates:

Temperature	0°C	5°C	10°C	15°C	20°C
ml CO_2/kg hr	2–3	4–6	8–10	10–16	14–25

Respiration rates of shredded cabbage are 13-20 mL CO_2/kg hr at 5°C.

To calculate heat production multiply mL CO_2/kg hr by 122 to get kcal/metric ton/day.

(*d*) Rates of Ethylene Production

Ethylene production rates are generally very low: <0.1 µL/kg·hr at 20°C (68°F), although higher rates have been reported for Chinese cabbage.

(*e*) Responses to Ethylene

Cabbages are sensitive to ethylene, which causes leaf abscission and leaf yellowing. Adequate ventilation during storage is important to maintain very low ethylene levels. Ethylene does not increase the disorder "black speck" or "pepper spot".

(*f*) Responses to Controlled Atmospheres (CA)

Some benefit to shelf-life can be obtained with low O_2 (2.5-5 per cent) and high CO_2 (2.5-6 per cent) atmospheres at temperatures of 0-5°C (32-41°F). CA storage will maintain color and flavor of cabbage, retard root and stem growth, and reduce leaf abscission. O_2 atmospheres below 2.5 per cent for round cabbage and 1 per cent for Chinese cabbage will cause fermentation, and CO_2 atmospheres >10 per cent will cause internal discoloration.

7. Postharvest Disorders, Diseases and their Control

Physiological Disorder

Black Speck

Black leaf speck (also called pepper spot, petiole spot, gomasho) is a disorder that consists of very small to moderate size discolored lesions on the midrib and veins of the leaves. The symptoms can occur after low temperatures in the field and by harvesting overmature heads, but are usually associated with transit and storage conditions. Low storage temperatures followed by warmer temperatures enhance development. Ethylene does not promote development of black speck in Chinese cabbage. Both round and Chinese cabbage cultivars vary widely in their susceptibility to this disorder. Storage with high CO_2 atmospheres (10 per cent) can reduce pepper spot development on round cabbage.

Cold Storage/Chilling Injury

It is purported to occur during storage at 0°C after 3 months or longer. The main symptom is midrib discoloration, especially on outer leaves. Cultivars differ greatly in their susceptibility to develop midrib discoloration.

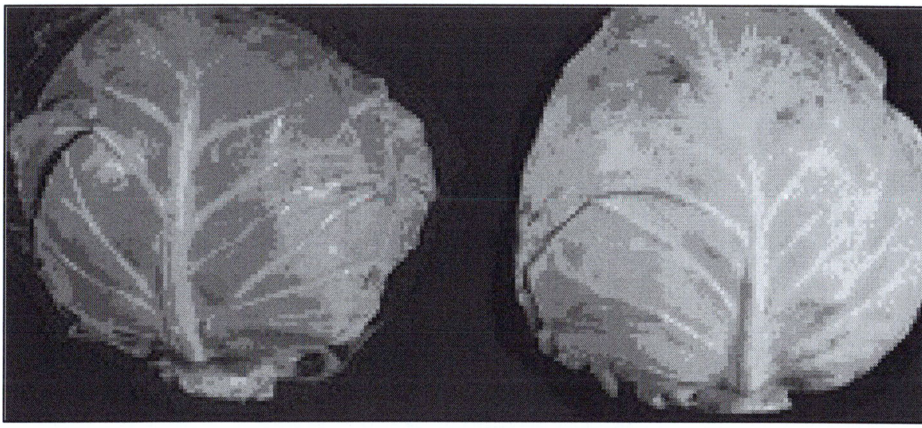

Ethylene-Induced Yellowing

Physical Injury

Breakage of the midribs often occurs during field packing and causes increased browning and susceptibility to decay. Outer midribs of overmature heads will crack easily.

Pathological Disorders

The most common decays found in stored cabbage are watery soft rot (*Sclerotinia*), gray mold rot (*Botrytis cinerea*), alternaria leaf spot (*Alternaria* spp.), and bacterial soft rot (caused by various bacterial species including *Erwinia, Pseudomonas, Xanthomonas*). Bacterial soft-rots result in a slimy breakdown of the infected tissue, and may follow fungal infections. Trimming outer leaves, rapid cooling and low temperature storage reduce development of these rots, although *Botrytis* and *Alternaria* will grow at low storage temperatures.

Important to Note

Fresh-cut or shredded cabbage pieces brown during storage and atmospheres of 3-5 per cent O_2 and 5-15 per cent CO_2 retard discoloration. Too low oxygen levels lead to fermentation and package blow-up, especially if product is not held below 5°C.

29

Cauliflower

Botanical Name: ***Brassica oleracea*** **var.** ***botrytis***

Family: Cruciferеae

1. Introduction

Cauliflower is the most popular vegetable among cole crops. It has small, thick stem, bearing whorl of leaves and branched tap root system. The main growing point develops into shortened shoot system whose apices make up the convex surface of curd, so the curd is a "prefloral fleshy apical meristem'. The eatable part, *i.e.* curd is generally white in colour and may be enclosed by inner leaves before its exposure. Curd colour varies with the variety and environment. It may be white, cream-white, yellow, green or red. Bihar, Uttar Pradesh, Orissa, West Bengal, Assam, Haryana and Maharashtra are major cauliflower-growing states. With the development of new varieties, it is now being grown in non-traditional areas-Andhra Pradesh, Tamil Nadu and Kerala.

2. Composition

Constituent	*Composition*
Moisture (%)	80
Protein (g)	5.9
Fat (g)	1.3
Minerals (g)	3.2
Fiber (g)	2.0
Carbohydrate (g)	7.6
Energy (K cal)	66
Calcium (mg)	625
Phosphrous (mg)	107
Iron (mg)	40
Carotene (mg)	30
Thiamine (mg)	0.04
Riboflavin (mg)	0.10
Niacin (mg)	1.0
Vitamin C (mg)	56

3. Suitable Varieties

September Maturity

Pusa Kunwari, Pant Gobi 3, Pusa Early,

October Maturity

Pusa Deepali, Pusa Katki,

November Maturity

Hisar 1, IIHR 101, Improved Japanese, Pant gobhi 2, PG, 26, Pusa Hybrid 2, Pusa Sharad

December Maturity

Pusa Himjyoti, Pusa Shubra, Pusa synthetic

Snow Ball Group

Dania, Pusa snow ball 1, Pusa snow ball K 1, Snow ball 16.

4. Maturity Indices

Cauliflowers are selected for size and compactness of the head or curd. Mature curds are at least 15 cm (6 inches) in diameter. Loose or protruding floral parts, creating a 'ricy' appearance, are a sign of overmaturity. Cauliflower is packaged after being closely trimmed into single layer cartons of 12 to 24 heads, with 12's most common.

Cauliflower is primarily marketed with closely trimmed leaves. Overwraps of perforated films should provide four to six 1/4-inch holes per head to allow adequate ventilation.

5. Quality Indices

A firm and compact head of white to cream white curds surrounded by a crown of well-trimmed, turgid green leaves. Additional quality indices are size, free from severe yellowing due to sunlight exposure, free from handling defects and decay, and an absence of 'riciness'.

6. Postharvest Management

Harvesting

Cauliflower is harvested when curds are compact, attain proper size and retain original colour. Delayed harvesting results in non-marketable loose and discoloured curds. The plant is cut well below the curd with a sharp cutting knife, sickle or khurpi. The trimming of the leaves depends upon the mode of packing and transportation. When packing is done in crates, most of the leaves are removed leaving small portion of the stalk close to the curd surface. While transporting in gunny bags, inner leaves covering the curd surface is left intact and outer leaves are removed. In loose transportation more number of leaves are retained and trimmed only after their unloading in the market.

Storage

(*a*) Optimum Temperature

0°C; 95-98 per cent R.H.

Storage of cauliflower is generally not recommended for more than 3 weeks for good visual and sensory quality. Wilting, browning, yellowing of leaves, and decay are likely to increase following storage beyond 3-4 weeks or at higher than recommended storage temperatures.

(*b*) Rates of Respiration/CO_2 Production

Temperature °C	*ml CO_2/kg hr*
0	8-9
5	10-11
10	16-18
15	21-25
20	37-42
25	43-48

To calculate heat production, multiply ml CO_2/kg hr by 122 to get kcal/metric ton/day.

(*c*) Rates of Ethylene Production

< 0.1 µl/kg· hr at 20°C

(d) Responses to Ethylene

Cauliflower is highly sensitive to exogenous ethylene. Discoloration of the curd and accelerated yellowing and detachment of wrapper leaf stalks will result from low levels of ethylene during distribution and short-term storage. Do not mix loads such as apples, melons and tomatoes with cauliflower.

(*e*) Responses to Controlled Atmosphere (CA)

Controlled or modified atmospheres offer moderate to little benefit to cauliflower. Injury from low O_2 (< 2 per cent) or elevated CO_2 (> 5 per cent) may not be visual and will only be evident after cooking. When the curds become grayish, extremely soft, and emit strong off-odor. Higher levels (>10 per cent) of CO_2 will induce this injury within 48 hours. Combined low O_2 and slightly elevated CO_2 levels (3-5 per cent) delay leaf yellowing and the onset of curd browning by a few days.

7. Postharvest Disorders, Diseases and their Control

Physiological Disorders

Cold Storage Injury is initiated at–0.8°C. Symptoms of cold storage injury or freezing injury include a watersoaked and grayish curd and watersoaked or wilted crown leaves. The curd will become brown and gelatinous in appearance following invasion by soft-rot bacteria.

Physical Injury

Harvesting should be done with great care to prevent damage to the highly sensitive turgid curds. Cauliflower should never be handled by the curd portion of the head. Cauliflower should never be allowed to roll or scuff across a harvest -conveyor belt, table, or other work surface.

Bruising

Bruising is very common and leads to rapid browning and decay when attention to careful harvest and handling practices are not followed.

Pathological Disorders

Diseases are an important source of postharvest loss, particularly in combination with rough handling and poor temperature control. A large list of bacterial and fungal pathogens cause postharvest losses in transit, storage, and to the consumer. Bacterial Soft-Rot (primarily *Erwinia* and

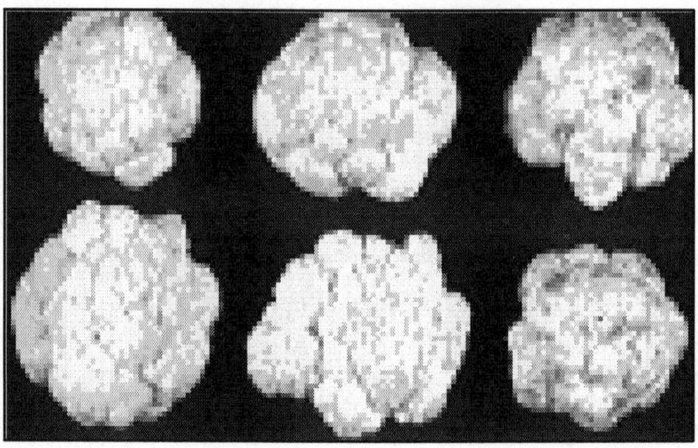

0°C 5°C 10°C

Discoloration (After 60 Days)

13d @7.5°C Air + 9.7 ppm C_2H_4

Ethylene Effects

(Physical damage followed by *Alternaria* spp.)

Mechanical Damage

Solar Yellowing

Pseudomonas), Black Spot (*Alternaria alternata.*), Grey Mould (*Botrytis cinerea*), and Cladosporium Rot are common disorders.

Important to Note

For fresh-cut applications, the sensitivity of cauliflower to improper modified atmosphere (See Responses to CA) demands very careful selection of packaging films and proper temperature management.

30

Carrot

Botanical Name: ***Daucus carota***

Family: Umbelliferae

1. Introduction

Carrot is grown all over India. It is taken raw as well as in cooked form. It is made into pickles and sweetmeat. Carrot juice is a rich source of carotene and is sometimes used for colouring butter and other foods. Black carrot is used for the preparation of a beverage called Kanji considered to be a good appetizer. Orange-coloured carrots are rich in carotene, a precursor of vitamin A and contain appreciable quantity of thiamine and riboflavin. The Asiatic types have more of anthocyanin pigments and less of carotene and may be less nutritive. Green caroot leaves are highly nutritive, rich in protein, minerals and vitamins and used as fooder and also for preparation of poultry feed. Main carrot growing states are Uttar Pradesh, Assam, Karnataka, Andhra Pradesh, Punjab and Haryana.

2. Composition (Per 100 g of Edible Portion *i.e.* carrot root)

Constituent	Composition
Carbohydrate (g)	10.6
Protein (g)	0.9
Fat (g)	0.2
Moisture (%)	86
Fiber (g)	1.2
Energy (kcal)	48
Minerals (g)	1.1
Iron (mg)	2.2
Carotene (mg)	1890
Thiamine (mg)	0.04
Riboflavin (mg)	0.02
Niacin (mg)	0.5
Vitamin C (mg)	3
Folic acid (mg)	15
Calcium (mg)	80
Phosphorus (mg)	30

3. Suitable Varieties

Pusa Kesar, Pusa Meghali, Sel 233, Chantenay, Early Nantes, Imperator, Nantes Hal Long, Pusa Yamdagini, Zeno.

4. Maturity Indices

- ☆ In practice, harvest decisions for carrots are based on several criteria depending on the market outlet or sales endpoint.
- ☆ Typically carrots are harvested at an immature state when the roots have achieved sufficient size to fill in the tip and develop a uniform taper.
- ☆ Length may be used as a maturity index for harvest timing of 'cut and peel' carrots to achieve a desired processing efficiency.

5. Quality Indices

There are many visual and organoleptic properties that differentiate the diverse varieties of carrots for fresh market and minimal processing. In general, Carrots should be:

- ✰ Firm (not flacid or limp)
- ✰ Straight with a uniform taper from 'shoulder' to 'tip'
- ✰ Bright orange
- ✰ There should be little residual "hairiness" from lateral roots
- ✰ No "green shoulders" or "green core" from exposure to sunlight during the growth phase.
- ✰ Low bitterness from terpenoid compounds
- ✰ High moisture content and high reducing sugars are most desirable for fresh consumption.

Quality defects include lack of firmness, non-uniform shape, roughness, poor color, splitting or cracking, green core, sunburn, and poor quality of tops or trimming.

6. Postharvest Management

Harvesting

Delay in harvesting of carrots makes roots pluffy and unfit for consumption. Early carrots for market are pulled out when partially developed. They are normally dug out with spade or khurpi when soil is sufficiently moist. The roots are trimmed and washed before sending to the market.

For washing, machines developed at AICRP (PHT) and PAU give better results.

Storage

(*a*) Optimum Temperature

0°C

Storage life at 0°C is typically:

Bunched: 10-14 days Immature roots: 4-6 weeks

Fresh-cut: 3-4 weeks Mature roots: 7-9 months

(Lightly processed)

Common storage conditions rarely achieve the optimum temperature for long- term storage to prevent decay, sprouting, and wilting. At storage temperatures of 3-5 °C, mature carrots can be stored with minimal decay for 3-5 months.

Common 'Cello-pack' carrots are typically immature and may be stored successfully for 2-3 weeks at 3-5°C. Bunched carrots are highly perishable due to the presence of the shoots (tops). Good quality is generally maintained only for 8-12 days, even with contact ice.

Lighlty processed (fresh-cut, cut and peel) carrots typically maintain quality of 2-3 weeks at 3-5°C.

(*b*) Optimum Relative Humidity

98-100 per cent; High relative humidity is essential to prevent dessication and loss of crispness. Free moisture from the washing process or unevaporated condensation, common with plastic bin-liners (and due to fluctuating temperatures) will promote decay.

(*c*) Rates of Respiration/CO_2 Production

Temperature °C	*ml CO_2/kg hr*	*Topped Bunched*
0	5–10	9–18
5	7–13	13–25
10	10–21	16–31
15	13–27	28–53
20	23–48	44–60
25	NA	NA

To calculate heat production multiply ml CO_2/kg hr by 122 to get kcal/metric ton/day. NA= not applicable

(*d*) Rates of Ethylene Production

>0.1µl/kg·hr at 20°C

(*e*) Responses to Ethylene

Exposure to ethylene will induce the development of bitter flavor due to isocoumarin formation. Exposure to as little as 0.5ppm exogenous ethylene will result in perceptible bitter flavor, within 2 weeks, at normal storage conditions. Thus, carrots should not be mixed with ethylene-producing commodities.

(*f*) Responses to Controlled Atmospheres (CA)

Controlled atmosphere is of limited use for carrots and does not extend postharvest life of carrots beyond that in air. CO_2 concentrations above 5 per cent have been shown to increase spoilage. Low oxygen concentrations, below 3 per cent, are not well tolerated and generally results in increased bacterial rot.

7. Postharvest Disorders, Diseases and their Control

Physiological and Physical Disorders

Intact Roots

Bruising, shatter-cracks and tip-breakage are signs of rough handling. Nantes-type carrots are particularly susceptible. *Sprouting* will continue as carrot roots develop new shoots after harvest. This is one reason low temperature postharvest management is critical. Common associated disorders include wilting, shriveling, or rubberiness due to dessication. *White Root* is a physiologic disorder due to suboptimal production conditions which results in patchy or streaks of low color on the carrot roots.

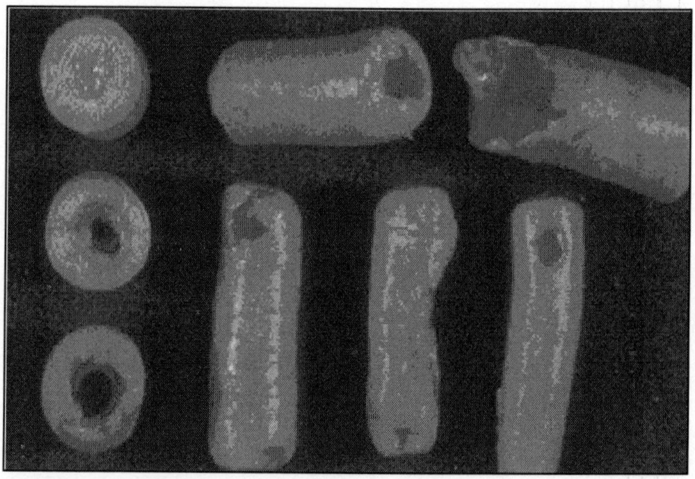

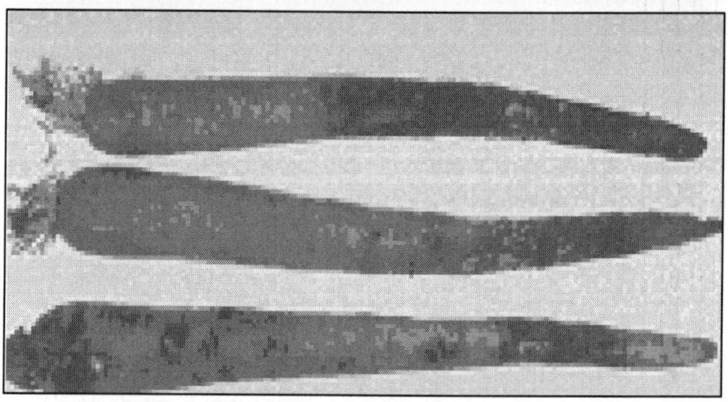

Bacterial Soft Rot Black Mold

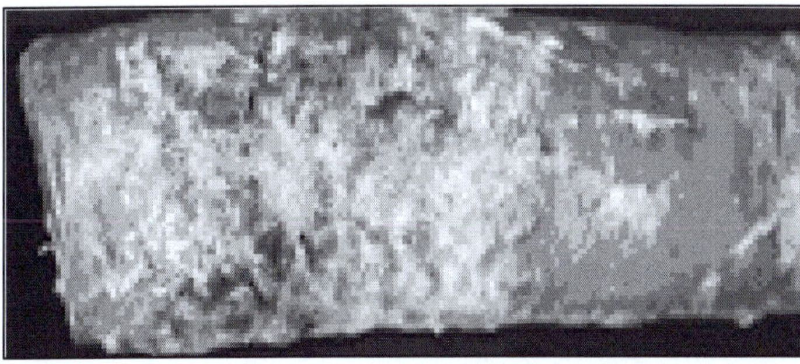

Gray Mold Rot

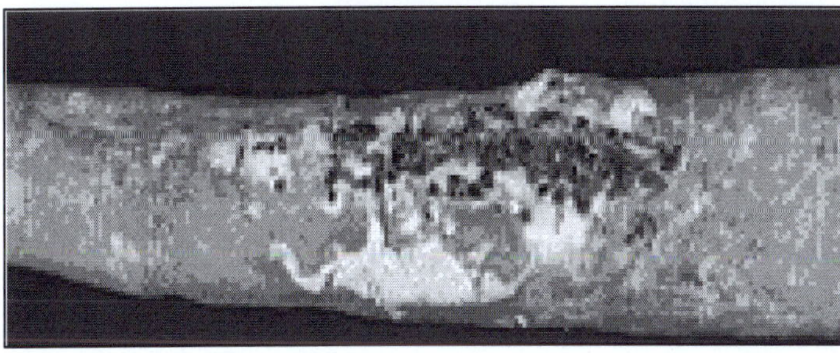

White Rot

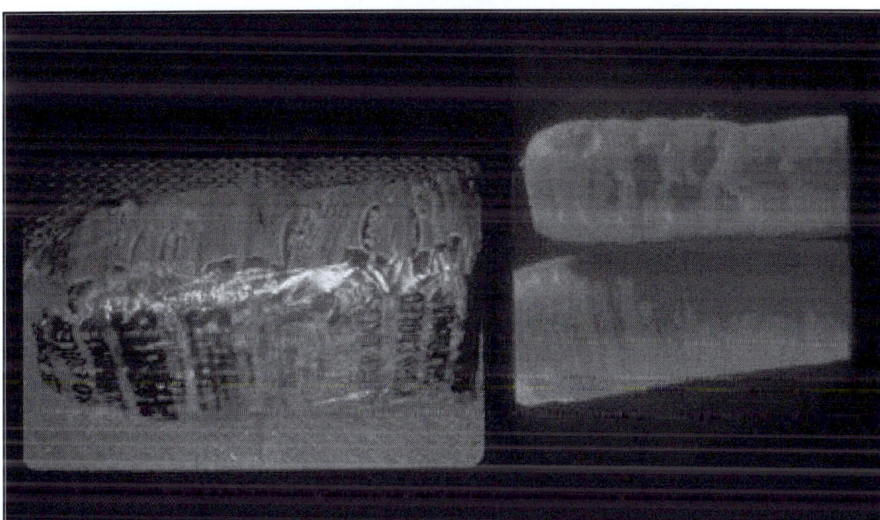

Pitting

Intact or Fresh-cut

Bitterness may be caused by preharvest stress (improper irrigation scheduling) or exposure to ethylene from ripening rooms or mixing with commodities such as apples. Freezing injury will likely result at temperatures of -1.2°C or lower. Frozen carrots generally exhibit an outer ring of water-soaked tissue, viewed in cross section, which blackens in 2-3 days.

Fresh-cut

White Blush, due to dehydration of cut or abrasion-peeled surfaces, has been a problem on fresh-cut carrots. Sharp cutting blades and residual free-moisture on the surface of the processed carrots will significantly delay the development of the disorder.

Pathological Disorders

The most prominent postharvest disease concerns are Gray Mold (*Botrytis rot)* Watery Rot (*Sclerotinia rot*), Rhizopus rot, Bacterial Soft Rot, induced by *Erwinia carotovora* ssp. *carotovora* and Sour Rot (*Geotrichum rot*). Proper handling and low temperature storage and transportation conditions are the best methods to minimize losses.

Important to Note

Rapid hydrocooling soon after harvest is strongly recommended.

31

Cucumber

Botanical Name: ***Cucumis sativus***

Family: Cucurbitiaceae

1. Introduction

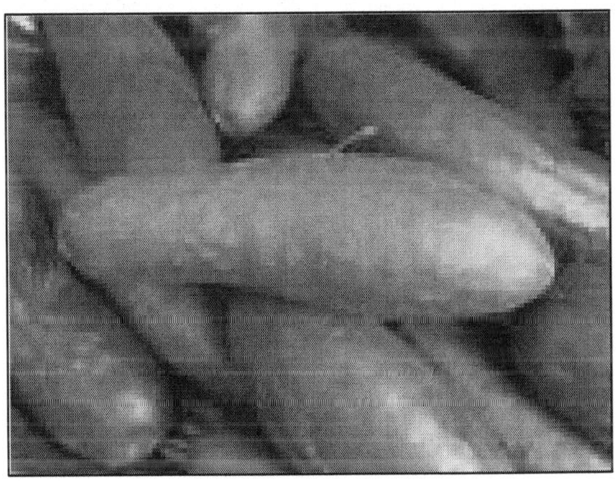

Cucumber is cultivated mostly for fresh consumption. India is considered to be the home of cucumber. It is an important salad crop cultivated both in north and south and lower as well as higher hills in India. Fruits varying in shape, size and colour contain 0.4 per cent protein, 2.5 per cent carbohydrates, 1.5 mg iron, and 2 mg of vitamin C in 100 g of fresh weight. Fruits are good for people suffering from constipation, jaundice and indigestion.

2. Composition (Per 100 g of Edible Portion)

Constituent	*Composition*
Moisture (%)	96.3
Protein (g)	0.4
Fat (g)	0.1
Minerals (g)	0.3
Fiber (g)	0.4
Carbohydate (g)	2.5
Energy k cal	13
Calcium (mg)	10
Phosphorous (mg)	25
Iron (mg)	0.6
Thiamine (mg)	0.02
Niacin (mg)	0.2
Folic acid (m g)	27.3
Vitamin C (mg)	7

3. Suitable Varieties

Himango, Japanese Long gree, Poinsett. Poona Khira, Pusa Sanyog, Sheetal.

4. Maturity Indices

Cucumbers are harvested at a range of developmental stages. Depending on cultivar and temperature, the time from flowering to harvest may be 55 to 60 days. Generally fruits are harvested at a slightly immature stage, near full size but before seeds fully enlarge and harden. Firmness and external glossiness are also indicators of a pre-maturity condition. At proper harvest maturity, a jelly like material has begun to form in the seed cavity.

5. Quality Indices

Table or slicing cucumber quality is primarily based on uniform shape, firmness and a dark green skin color. Additional quality indices are size, freedom from growth or handling defects, freedom from decay, and an absence of yellowing.

6. Postharvest Management

Harvesting

The cucumber should be harvested at 2 days interval. Picking is mostly done manually, however for, quality produce the cucumbers are trimmed with clippers and washed thoroughly with machines developed at PAU, Ludhiana.

Storage

(*a*) Optimum Temperature and Relative Humidity

10–12.5°C; 95 per cent R.H.

Storage of cucumber is generally less than 14 days as visual and sensory quality deteriorate rapidly. Shriveling, yellowing, and decay are likely to increase following storage beyond two weeks, especially after removal to typical retail conditions. Short term storage or transit temperatures below this range (such as 7.2°C) are commonly used but will result in chilling injury after 2-3 days.

(*b*) Chilling Injury

Cucumbers are chilling sensitive at temperatures below 10°C if held for more than a day to 3 days depending on temperature and cultivar. Consequences of chilling injury are water-soaked areas, pitting and accelerated decay. Chilling injury is cumulative and may be initiated in the field prior to harvest. Cucumber varieties vary considerably in their susceptibility to chilling injury.

(*c*) Rates of Respiration/CO_2 Production

Temperature	10°C	15°C	20°C	25°C
ml CO_2/kg hr	12–15	12–17	7–24	10–26

Respiration varies widely above 10°C due to different stages of maturity. Less mature cucumbers have higher respiration rates. To calculate heat production, multiply ml CO_2/kg·hr by 122 to get kcal/metric ton/day.

(*d*) Rates of Ethylene Production

0.1–1.0µl/kg·hr at 20°C

(*e*) Responses to Ethylene

Cucumbers are highly sensitive to exogenous ethylene. Accelerated yellowing and decay will result from low levels (1-5ppm) of ethylene during distribution and short-term storage. Do not mix commodities such as bananas, melons and tomatoes with cucumber.

(f) Responses to Controlled Atmospheres (CA)

Controlled or modified atmosphere storage or shipping offer moderate to little benefit to cucumber quality maintainence. Low O_2 levels (3-5 per cent) delay yellowing and the onset of decay by a few days. Cucumber tolerates elevated CO_2 up (CA) to 10 per cent but storage life is not extended beyond the benefit of reduced levels of O_2.

7. Postharvest Disorders, Diseases and their Control

Physiological Disorders

Cold Storage Injury

It is initiated at –0.5°C. Symptoms of include a watersoaked pulp becoming brown and gelatinous in appearance over time.

Physical Injury

Harvesting should be done by cutting free of the vine rather than by tearing. "Pulled end" is a quality defect used in establishing grade quality.

Chilling Injury

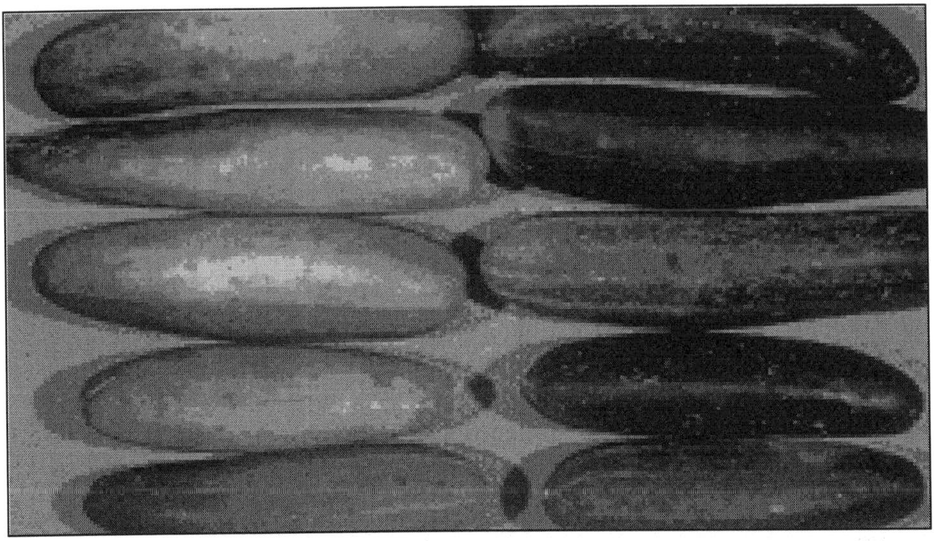

C_2H_4 + 12.5°C Control

Ethylene Damage

Fusarium Decay

Bruising and compression injury are very common when attention to careful harvest and handling practices are not followed.

Pathological Disorders

Diseases are an important source of postharvest loss, particularly in combination with chilling stress. A large list of bacterial and fungal pathogens cause postharvest losses in transit, storage, and to the consumer. *Alternaria* spp., *Didymella* Black Rot, *Pythium* Cottony Leak, and *Rhizopus* Soft Rot are common disorders.

Important to Note

Cucumbers are often treated with approved waxes or oils to reduce water loss, reduce abrasion injury and enhance appearance.

Yellowing during the postharvest period is a very common defect. Harvesting fruit at an advanced stage of development, exposure to ethylene, or storage at too high temperature all cause yellowing.

32

Brinjal

Botanical Name: ***Solanum melongena*** **L.**

Family: Solanaceae

1. Introduction

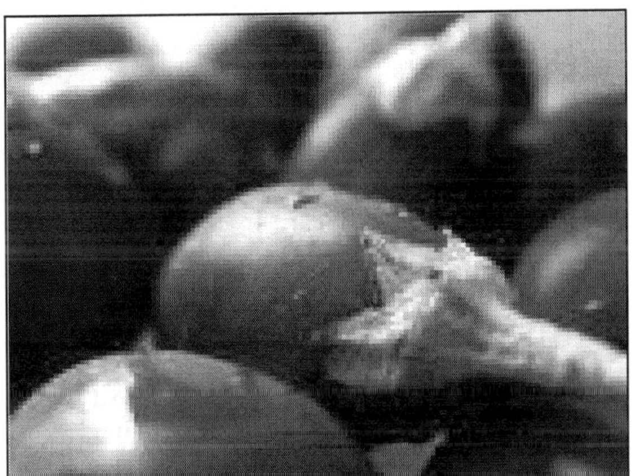

Brinjal is one of the most common, popular and principal vegetable crops grown in India and other parts of the world. It can be grown in almost all parts of India except higher altitudes,all the year round. A number of cultivars are grown throughout the country depending upon the yield, consumer's preference about the colour, size and shape of the various cultivars. The brinjals is of much importance in the warm areas of fareast, being grown extensively in India, Bangladesh, Pakistan, China and the Phillipines. It is also popular in France, Italy and the United States. It is highly productive and usually finds its place as the poor man's crop. In India, it is being consumed as a cooked vegetable in various ways. The crop is extremely variable in India.

2. Composition (Per 100g of Edible Portion)

Constituent	Composition
Moisture (%)	92.7
Protein (g)	1.4
Fat (g)	0.3
Minerals (g)	0.3
Fiber (g)	1.3
Carbohydrates(g)	4.0
Calcium (mg)	18
Magnesium (mg)	16
Oxalic acid (mg)	18
Phosphorus (mg)	47
Iron (mg)	0.9
Sodium (mg)	3.0
Potassium (mg)	2.0
Copper(mg)	0.17
Sulphur (mg)	44.0
Chlorine (mg)	52.0
Vitamin A (IU)	124
Thiamine (mg)	0.04
Riboflavin (mg)	0.11
Nictoninic acid (mg)	0.09
Vitamin C (mg)	12.0

3. Suitable Cultivars

Pusa purple long, Pusa purple cluster, Pusa Kranti, Pusa Anmol, Pusa purple round, Arka Sheel, Arka Shirish, Arka Kusumukar, Arka Navneet, Punjab Barsati, Vijay Hybrid, Long 13, Round 14, Suphal

4. Maturity Indices

Brinjal fruits are harvested at a range of developmental stages. Depending on cultivar and temperature, the time from flowering to harvest may be 10 to 40 days. Generally fruit are harvested immature before seeds begin to significantly enlarge and harden. Firmness and external glossiness are also indicators of a pre-maturity condition. Brinjal fruit become pithy and bitter as they reach an overmature condition.

5. Quality Indices

The diversity of brinjal types being marketed has increased greatly in recent years. Standard brinjal quality is primarily based on uniform egg to globular shape, firmness and a dark purple skin color. Distinction among grades is based solely on size, external appearances, and firmness.

6. Postharvest Management

Harvesting

The fruits are harvested with stalk at joints where they are attached to the branches. Some cultivars have soft joints and are easy to harvest while in others help of sharp knife is required. Care should be taken to avoid injury to the branches.

Storage

(*a*) Optimum Temperature and Relative Humidity

10–12°C; 90-95 per cent R.H.

Storage of brinjal is generally less than 14 days as visual and sensory qualities deteriorate rapidly. Decay is likely to increase following storage beyond two weeks, especially after removal to typical retail conditions. Short term storage or transit temperatures below this range are used often to reduce weight loss, but will result in chilling injury after several days.

(*b*) Rates of Respiration/CO_2 Production

Temperature	12.5°C
ml CO_2/kg hr	30-39

To calculate heat production, multiply ml CO_2/kg·hr by 440 to get BTU/ton/day or by 122 to get kcal/metric ton/day.

(*c*) Rates of Ethylene Production

0.1–0.7μl/kg·hr at 12.5°C

(*d*) Responses to Ethylene

Brinjal fruits have a moderate to high sensitivity to exogenous ethylene. Calyx abscission and increased deterioration, particularly browning, may be a problem if brinjal is exposed to >1ppm ethylene during distribution and short-term storage.

(*e*) Responses to Controlled Atmospheres (CA)

Controlled or modified atmosphere storage or shipping offers little benefit to brinjal quality maintenance. Low O_2 levels (3-5 per cent) delay deterioration and the onset of decay by a few days. Brinjal tolerates up to 10 per cent CO_2 but storage life is not extended beyond the benefit of reduced levels of O_2.

7. Postharvest Disorders, Diseases and their Control

Physiological Disorders

Chilling Injury

Brinjal fruits are chilling sensitive at temperatures below 10°C. At 5°C chilling injury will occur in 6-8 days. Consequences of chilling injury are pitting, surface bronzing, and browning of seeds and pulp tissue. Accelerated decay by *Alternaria* spp. is common in chilling stressed fruit. Chilling injury is cumulative and may be initiated in the field prior to harvest.

Freezing Injury

Freezing injury will be initiated at–0.8°C, depending on the soluble solids content. Symptoms of freezing injury include a watersoaked pulp becoming brown and desiccated in appearance over time.

Physical Injury

Harvesting should be done by cutting the calyx-stem free from the plant rather than by tearing. Cotton gloves are often used.

Bruising and Compression Injury

Bruising and compression injury are very common when attention to careful harvest and handling practices are not followed. Brinjal cannot withstand stacking in bulk containers.

Pathological Disorders

Diseases are an important source of postharvest loss, particularly in combination with chilling stress. Common fungal pathogens are *Alternaria* (Black Mold Rot), *Botrytis* (Gray Mold Rot), *Rhizopus* (Hairy Rot), and *Phomopsis* Rot.

Chilling Injury

Pitting

Important to Note

Rapid cooling, primarily to reduce water loss, soon after harvest is essential for optimal postharvest keeping quality. The precooling endpoint is typically 10°C. Forced-air cooling is the most effective practice. Room cooling after washing or hydrocooling is the most common practice. Moistened paper or waxed cartons are often used to reduce water loss.

Chilling injury and water loss can be reduced by storing of brinjal in polyethylene bags or polymeric film overwraps. Increased decay from *Botrytis* is a potential risk of this practice.

Solar Yellowing

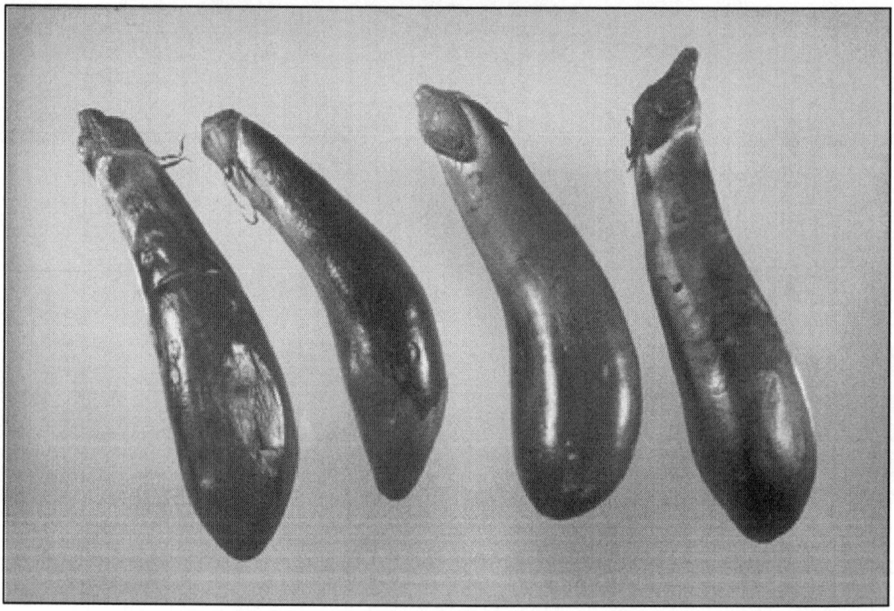

Sun Scald

33

Garlic

Botanical Name: ***Allium sativum***

Family: Liliaceae

1. Introduction

Garlic is an important crop. It is mainly used for flavouring and seasoning vegetables and meat dishes. Madhya Pradesh, Gujarat, Orissa, Maharashtra, Uttar Pradesh and Rajasthan are major garlic-growing states. More than 50 per cent production, however, comes from Madhya Pradesh and Gujarat only. The area, production and productivity and per capita availability of garlic have improved considerably. Indore and Mandsour in Madhya Pradesh, Angul in Orissa, Junnar Taluka in Maharashtra, Mainpuri, Etah and Etawa in Uttar Pradesh, Sikar, Jhunjhun, Ajmer and Udaipur in Rajasthan are major garlic-growing pockets. In India smaller-cloved garlic varieties are being grown in plains.

Recently excess larger-cloved garlic cultivation has been introduced in north Indian plains and Madhya Pradesh. Larger-cloved garlic varieties are also cultivated in Himachal Pradesh and Nilgiri hills.

2. Composition (Per 100g Edible Portion)

Constituent	*Composition*	
	Fresh Peeled Garlic Cloves (Bulb Lets)	*Dehydrated Garlic Powder*
Moisture (per cent)	62.80	5.20
Protein (g) per cent	6.30	17.50
Fat (g)	0.10	0.60
Mineral matter (g)	1.00	3.20
Fiber (g)	0.80	1.90
Carbohydrates (g)	29.00	71.40
Calcium(mg)	0.03	0.10
Phosphorus(mg)	0.31	0.42
Potassium (mg)	–	1.10
Iron (mg)	0.001	0.004
Niacin (mg)	–	0.70
Sodium (mg)	–	0.01
Vitamin A (I.U.)	0.0	175.00
Nocotinic acid (mg)	0.40	–
Vitamin C (mg)	13.00	12.00
Vitamin B (mg)	–	0.68
Vitamin B_2 (mg)	–	0.08

3. Suitable Varieties

Godavari, Sweta, HGI, HG6, Pusa Sel 10, LCG, 1ARU 52, Agrifound white (G41), Yamuna safed and Yamuna Safed2, G 282 and Agrifound Parvati.

4. Maturity Indices

Garlic can be harvested at different stages of development for speciality markets, but most garlic is harvested when the bulbs are well mature. Harvest occurs after the tops have fallen and are very dry.

5. Quality Indices

High quality garlic bulbs are clean, white (or other colors typical of the variety), and well cured (dried neck and outer skins). The cloves should be firm to the touch. Cloves from mature bulbs should have a high dry weight and soluble solids content (>35 per cent in both cases).

6. Postharvest Management

Harvesting

Garlic is mostly harvested manually by hand digger. In some regions where the soil is loose, the plants are pulled out by hand or pulling is done by loosening the soil with country plough. Bulbs are taken out along with their tops and windrowed gathering several rows in each row.

Storage

(*a*) Optimum Temperature

–1°C to 0°C The variety of garlic affects potential storage life, and the recommended conditions for commercial storage depend on the expected storage period. Garlic can be kept in good condition for 1-2 months at ambient temperatures (20°-30°C) under low relative humidity (<75 per cent). However under these conditions, bulbs will eventually become soft, spongy and shriveled due to water loss. For long-term storage, garlic is best maintained at temperatures of -1°C to 0°C with low relative humidity (60-70 per cent). Good airflow is also necessary to prevent any moisture accumulation. Under these conditions garlic can be stored for more than 9 months.

Garlic will eventually lose dormancy, signaled by internal development of the sprout. This occurs most rapidly at intermediate storage temperatures of 5° 18°C. Garlic odor is easily transferred to other products and should be stored separately. High humidity in the storages will favor mold growth and rooting. Mold growth can also be problematic if the garlic has not been well cured before storing.

(*b*) Optimum Relative Humidity

60 to 70 per cent.

(*c*) Rates of Respiration/CO_2 Production

Temperature	0°C	5°C	10°C	15°C	20°C
ml CO_2/kg hr					
Intact bulbs	2–6	4–12	6–18	7–15	7–13
Fresh peeled cloves	12	15–20	35–50		

To calculate heat production multiply ml CO_2/kg hr by 122 to get kcal/metric ton/day.

(*d*) Rates of Ethylene Production

Garlic produces only very low amounts of ethylene (<0.1 μ/kg·hr)

(*e*) Responses to Ethylene

Not sensitive to ethylene exposure.

(*f*) Responses to Controlled Atmospheres (CA)

Atmospheres with high CO_2 (5-15 per cent) are beneficial in retarding sprout development and decay during storage at 0-5°C. Low O_2 alone (0.5 per cent) did not retard sprout development of 'California Late' garlic stored up to 6 months at 0°C. Atmospheres with 15 per cent CO_2 may result in some yellow translucent discoloration occurring on some cloves after about 6 months

7. Postharvest Disorders, Diseases and their Control

Physiological Disorders

Freeze Injury

Due to its high solids content, garlic freezes at temperatures below -1°C.

Waxy Breakdown

Waxy breakdown is a physiological disorder that affects garlic during latter stages of growth and is often associated with periods of high temperature near harvest. Early symptoms are small, light yellow areas in the clove flesh that darken to yellow or amber with time. Finally the clove is translucent, sticky and waxy, but the outer dry skins are not usually affected. Waxy breakdown is commonly found in stored and shipped garlic but rarely in the field. Low oxygen levels and inadequate ventilation during handling and storage may also contribute to development of waxy breakdown.

Pathological Disorders

Penicillium Rots

Penicillium rots (*Pencillium corymbiferum* and other spp.) are common problems in stored garlic. Affected garlic bulbs may show little external evidence until decay is advanced. Affected bulbs are light in weight and the individual cloves are soft and spongy and powdery dry. In an advanced stage of decay, the cloves break down in a green or gray powdery mass. Low humidity in storage retards rot development. Less common storage decay problems include *Fusarium basal rot* (*Fusarium oxysporum* cepae) which infects the stem plate and causes shattering of the cloves, *dry rot* due to *Botrytis allii*, and bacterial rots (*Erwinia* spp., *Pseudomonas* spp.).

Important to Note

To control sprout development and lengthen the storage period, garlic may be treated with preharvest applications of sprout inhibitors (*i.e.*, maleic

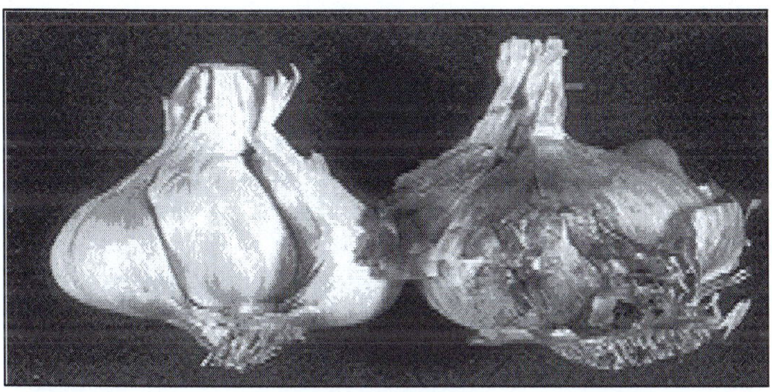

Black Mold

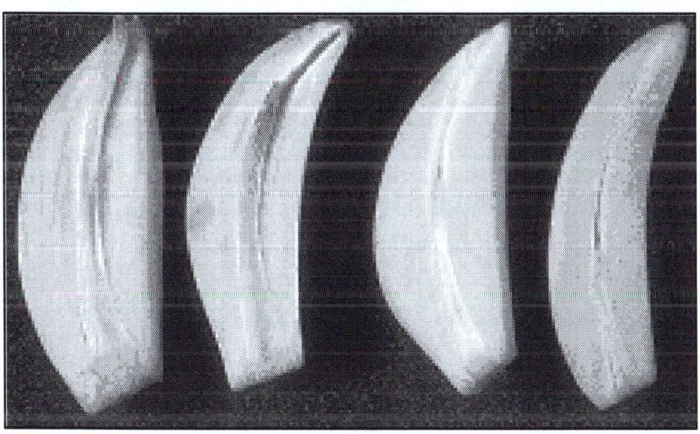

Sprout Development

hydrazide) or be irradiated after harvest. Outer cloves of bulbs are easily damaged during mechanical harvest and these damaged areas discolor and decay during storage. Therefore high quality garlic for the fresh market is usually harvested manually to avoid mechanical damage.

Curing garlic is the process by which the outer leaf sheaths and neck tissues of the bulb are dried. Warm temperatures, low relative humidity, and good airflow are conditions needed for efficient curing. Under favorable climatic conditions in California, the garlic is usually cured in the field. Curing is essential to obtain maximize storage life and have minimal decay.

Garlic flavor is due to the formation of organosulfur compounds when the main odorless precursor alliin is converted by the enzyme alliinase to allicin and other flavor compounds. This occurs at low rates unless the garlic cloves are crushed or damaged. Alliin content decreases during storage of garlic bulbs, but the effect of time, storage temperatures and atmospheres has not yet been well documented.

34

Okra

Botanical Name: *Abelmoschus esculentum* **L.**

Family: Malvaceae

1. Introduction

Okra (bhindi) is an annual vegetable crop grown from seed in tropical and subtropical parts of the world. Its tender green fruits are used as a vegetable and are generally marketed in the fresh state.

2. Composition (Per 100 g of Edible Portion)

Constituent	*Composition*
Moisture (%)	89.6
Carbohydrates(g)	6.4
Protein(g)	1.9
Fat (g)	0.2
Fiber (g)	1.2

Constituent	Composition
Minerals (g)	0.7
Calcium (mg)	66
Magnesium (mg)	43
Oxalic acid (mg)	8
Phosphorus (mg)	56
Iron (mg)	1.5
Sodium (mg)	6.9
Potassium (mg)	103
Copper (mg)	0.19
Sulphur(mg)	30
Vitamin A (IU)	88
Thiamine (mg)	0.07
Riboflavin (mg)	0.10
Nictonic acid (mg)	0.60
Vitamin C (mg)	13

3. Suitable Varieties

Pusa Makhmali, Punjab No. 13, Pusa Sawani, IHR 20-31, Red Bhindi, Punjab Padmini

4. Maturity Indices

Okra pods are immature fruits and are harvested when they are very rapidly growing. Harvesting typically occurs 3 to 7 days after flowering. Okra should be harvested when the fruit is bright green, the pod is fleshy and seeds are small. After that period, the pod becomes pithy and tough, and the green color and mucilage content decrease.

5. Quality Indices

Okra pods should be tender and not fibrous, and have a color typical of the cultivar (generally bright green). The pods should be well formed and straight, have a fresh appearance and not show signs of dehydration. Pods are packed based on length with Fancy, Choice and Jumbo designations for size categories. Okra should be free of defects such as leaves, stems, broken pods, insect damage, and mechanical injury. The tender pods are easily damaged during harvest, especially on the ridges and this leads to unsightly brown and black discoloration. Quality losses that occur during marketing are often associated with mechanical damage, water loss, chilling injury, and decay.

6. Postharvest Management

Harvesting

Early harvesting gives lower yields of tender fruits with shorter shelf life.

In general harvesting on every alternate day is advisable. A low cost hand glove or cloth bag should be used to protect fingers. Harvesting in the morning is convenient. For distant markets, harvesting during late late evening and transportation during night is also advisable. The fruits are graded. For processing industry and fresh fruit export 6-8 cm long fruits are sorted out. Longer fruits are used for the fresh market.

Pre Cooling

For local market fruits are cooled (preferably) and packed in jute bags or baskets, covered or sewed and then water sprinkled over it. This helps in cooling and keeping turgidity of the fruits intact and saves product from blemishes, bruises and blackening.

Storage

(*a*) Optimum Storage Temperature

7-10°C

Very good quality can be maintained up to 7 to 10 days at these temperatures. If stored at higher temperatures, the pods lose quality due to dehydration, yellowing and decay. When stored at lower than recommended temperatures, chilling injury will be induced (see physiological disorders). Chilling symptoms include surface discoloration, pitting and decay. Okra can be successfully hydrocooled or forced-air cooled.

(*b*) Optimum Relative Humidity

Weight loss is very high in immature okra pods and cultivars may vary in rate of water loss. A very high relative humidity (95-100 per cent) is needed to retard dehydration, pod toughening, and loss of fresh appearance.

(*c*) Rates of Respiration/CO_2 Production

Okra pods have very high respiration rates.

Temperature	5°C	10°C	15°C	20°C
ml CO_2>/kg hr	27–30	43–47	69–72	124–137

To calculate heat production multiply mL CO_2/kg hr by 122 to get kcal/metric ton/day.

(*d*) Rates of Ethylene Production and Responses to Ethlylene

Okra pods have low ethylene production rates (<0.5 µL/kg·hr at 10°C). Exposure to ethylene reduces shelf-life by increasing pod yellowing.

(*e*) Responses to Controlled Atmospheres (CA)

Okra is not stored in modified atmospheres commercially. At recommended storage temperatures, CO_2 concentrations of 4-10 per cent can help to maintain green color and reduce discoloration and decay on damaged pods. CO_2 concentrations higher than 10 per cent can lead to off flavors. Low O_2 concentrations (3-5 per cent) reduce respiration rates and may also be beneficial.

7. Postharvest Disorders, Diseases and their Management

Physiological Disorders

Chilling Injury

The typical symptoms of chilling injury in okra are discoloration, pitting, water-soaked lesions and increased decay (especially after removal to warmer temperatures, as during marketing). Different cultivars may differ in their susceptibility to chilling injury. Calcium dips and modified atmospheres have been reported to reduce chilling symptoms.

Freeze Damage

Occurs at temperatures of -1.8°C or below.

Pathological Disorders

Decay on okra can be due to various common bacterial and fungal organisms, but chilling and injury-enhanced rots are probably the most common causes of loss. *Rhizopus, Geotrichum* and *Rhizoctonia* fungal rots as well as bacterial decays due to *Pseudomonas* sp. have been reported to cause postharvest losses.

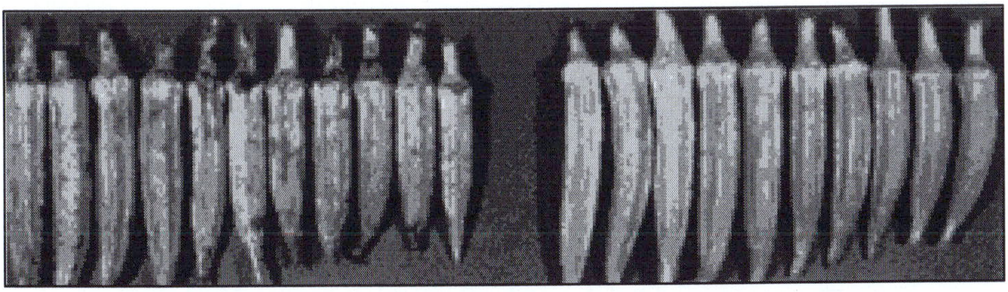

Air Control **Air + 5 per cent Co_2**

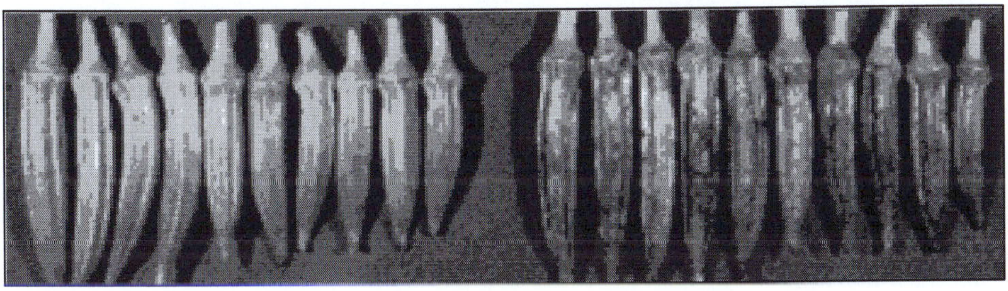

Air + 10 per cent Co_2 **Air+15 per cent Co_2**

Chilling Injury

After 8 days in indicated atmosphere at 5C+ 3 days in air at 20°C
Intermediate carbon dioxide atmosphere reduce chilling injury of okra

35

Onion

Botanical Name: ***Allium cepa***

Family: Liliaceae

1. Introduction

Among the bulb crops, onion and garlic are the most important vegetable crops grown in India. The onion is grown from ancient times in India. It is one of the few versatile vegetable crops that can be kept for a fairly long period and cab safely withstand the hazards of rought handling including long distance transport. Onion is an important crop in all continents with world production of about 25 million tones and is commercially cultivated in a little over hundred countries of the world. However, about three-fourth of global production is accounted for Spain, Turkey, Brazil, Iran etc. India's share in the world production is about 11 per cent and occupies second place.

2. Composition

Constituent	*Composition*
Moisture (%)	86.8
Carbohydrates (g)	11.0
Protein (g)	1.2
Fiber (g)	0.6
Minerals (g)	0.4
Thiamine (mg)	0.08
Vitamin C (mg)	11
Calcium (mg)	180
Phosphorus (mg)	50
Iron (mg)	0.7
Nicotinic acid (mg)	0.4
Riboflavin (mg)	0.01

3. Suitable Varieties

Red cultivars, white cultivars, yellow cultivars, Bermuda yellow, Early grano, Pusa red, Pusa Ratnar, Pusa white flat, Arka niketan, Arka kalyan

4. Maturity Indices

- ✰ Indicated when approximately 10 to 20 per cent of tops have fallen over
- ✰ Conversion from active growth to dormancy accelerated by undercutting bulbs 1 to 2 inches
- ✰ "Field-dry" maturity is indicated when bulb neck is completely dry to the touch and not slippery. Typically reached at 5-8 per cent weight loss following harvest.

5. Quality Indices

- ✰ Mature neck and scales
- ✰ Firmness
- ✰ Diameter (Bulb size)
- ✰ Absence of decay, insect damage, sunscald, greening, sprouting, freezing injury, bruising, and other defects
- ✰ Degree of pungency

6. Postharvest Management

Harvesting

Onion is ready for harvesting in 3-5 months and 2-3 months after transplanting for dry and green onions respectively. Green onions are harvested when they are just ready for earthing. Plants are uprooted with hand and their roots are cut. They are washed and bundled as per the market requirement.

Storage

(*a*) Optimum Temperature

Curing

Field curing when temperatures are at least 24°C or exposure for 12 hrs. to 30 to 45°C for forced air-curing.

Storage

Mild onions: Typically 0.5 to 1 month at 0°C

Pungent Onions: Typically up to 6 to 9 months at 0°C depending on the cultivar

(*b*) Optimum Relative Humidity

Curing

75 to 80 per cent for best scale color development

Storage

65 to 70 per cent with adequate air circulation ($1m^3/min/m^3$ of onion)

(*c*) Rates of Respiration/CO_2 Production

- ☆ Whole Onions- 3-4 ml/kg·hr @ 0-5°C; 27-29 ml/kg/hr @ 25-27°C. Storage between 5-25°C favors sprouting and is not recommended for extended periods.
- ☆ Diced Onions- 40-60 ml/kg·hr @ 0-5°C.

To calculate heat production multiply ml CO_2/kg·hr by 122 to get kcal/metric ton/day.

(*d*) Rates of Ethylene Production

Whole Onions: < 0.1 µl/kg·hr at 0-5°C

Diced Onions NA

(*e*) Responses to Ethylene

Ethylene may encourage sprouting and growth of decay-causing fungi.

(*f*) Responses to Controlled Atmospheres (CA)

No commercial benefit has been identified for varieties with long storage potential. Onions are damaged by < 1 per cent O_2 and 10 per cent CO_2. There is some commercial use of CA (3 per cent $O_2$and 5-7 per cent CO_2) for sweet onion varieties (short storage potential). Diced onions benefit from CA conditions of 1.5 per cent O_2 and 10 per cent CO_2.

7. Postharvest Disorders, Diseases and their Control

Physiological Disorders

Freezing Injury

Soft water-soaked scales rapidly decay from subsequent microbial growth.

Translucent Scales

Resembles freezing injury and is prevented by prompt cold storage following curing; 3-4 week delay in cold storage increases risk significantly.

Greening

Exposure to light following curing causes green-coloration of outer scales.

Ammonia Injury

Brown-black blotches result from ammonia gas leakage during storage.

Pathological Disorders

Botrytis Neck Rot

Watery-decay initiates at neck area and moves downward through entire bulb. Light gray to gray fungal growth is generally visible at neck infection and on outer scales. Proper drying and curing of onion essentially prevents this storage disorder. Storage conditions (as above) should be maintained to prevent condensation from forming on the bulbs.

Black Mold

Black discoloration and shriveling at neck and on outer scales caused by the fungus *Aspergillus niger*. Often associated with bruising and leads to bacterial soft rot. Low temperature storage will delay growth of fungus following field or handling infestation but growth will resume above 15°C.

Blue Mold

Watery soft rot of neck and outer scales followed by the appearance of green-blue mold (occasionally yellow-green) spores of the fungus Penicillium. Minimize bruising and other mechanical injuries, sunscald, and freezing injury.

Bacterial Rots/Soft Rot

Water-soaked, foul-smelling, viscous liquidy rot caused by *Erwinia carotovora* subsp. *carotovora*.

Slippery Skin

Generally visible only at neck area and upon cutting to expose inner scales. Scales have a watery-cooked appearance.

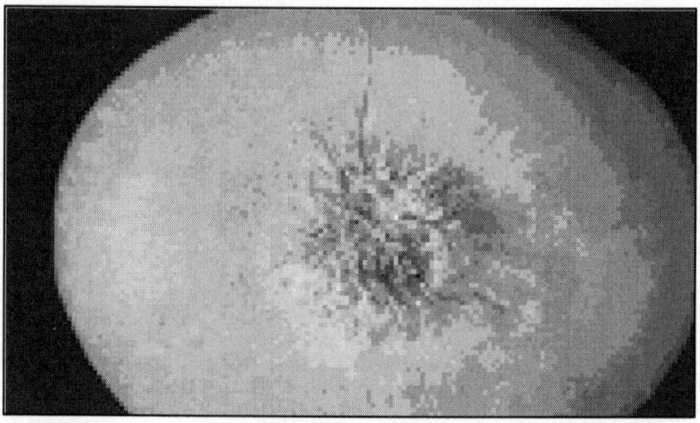

Fusarium Basal Rot

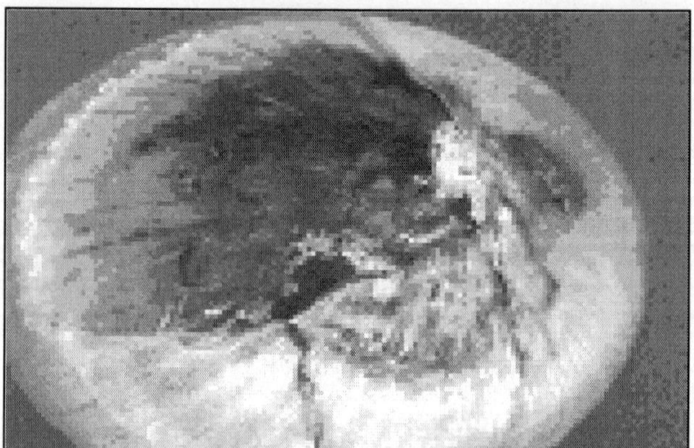

Fusarium Bulb Rot

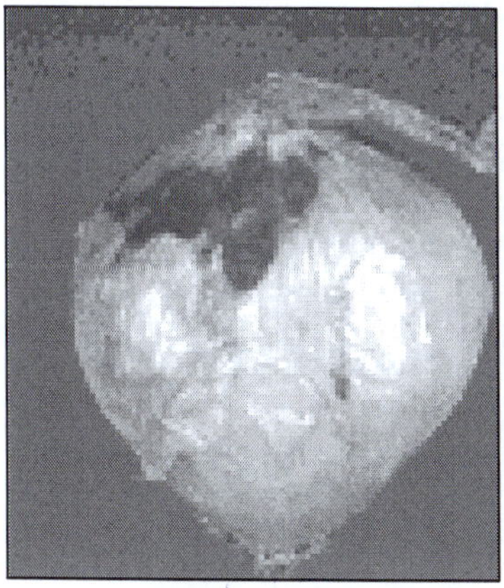

Grey Mold

Neck Rot

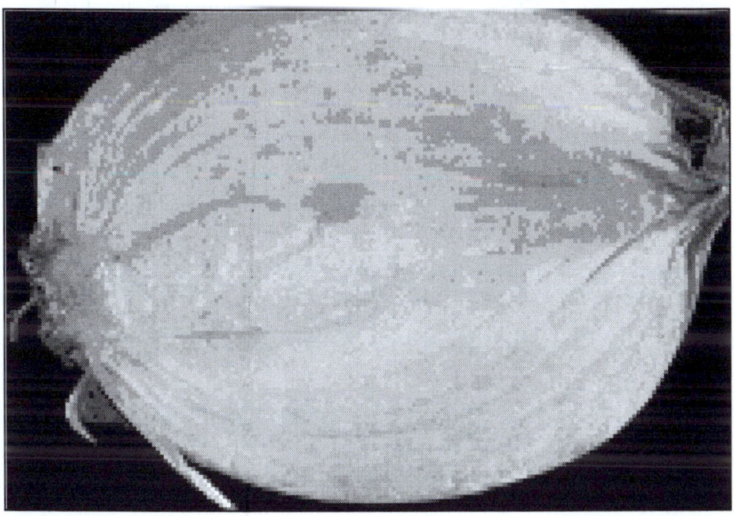

Soft Rot

Sour Skin

Slimy, yellow brown decay generally limited to inner scales which give off a sour odor when exposed.

General Bacterial Rot Control

1. Harvest only at full maturity
2. Proper drying and curing
3. Minimizing bruising and scraping;
4. Maintaining proper storage conditions (as above) to prevent condensation from forming on the bulbs.

Translucent Scales

Important to Note

Onions are both storage-odor sources for other commodities, such as apples, celery and pears, and storage-odor absorbers from commodities such as apples.

36

Pea

Botanical Name: ***Pisum sativum L.***

Family: Leguminaceae

1. Introduction

In India, it is grown as winter vegetable in the plains of North India and summer vegetable in the hills. Maximum cultivation of pea is in Uttar Pradesh which accounts for about 60 per cent area under this crops, followed by Bihar and Madhya Pradesh. Pea is highly nutritive containing high percentage of digestible protein, along with carbohydrates and vitamins. It is also very rich in minerals. Being a nitrogen-fixing legume, its value as a green manure crop has long been recognized. The interest in pea as a soil building crop will increase day by day as the chemical fertilizers are becoming less available and more expensive. Edible-pod peas include both Oriental or Asian flat type pods, harvested when the seeds are very small and immature, and the Snap or Sugar Snap Pea which resemble a typical fresh garden pea but with smaller seeds

2. Composition (Per 100g of Edible Portion)

Constituent	Composition	Constituent	Composition
Moisture (%)	72.0	Sulphur (mg)	95
Protein(g)	7.2	Iron (mg)	1.5
Fat(g)	0.1	Thiamine (mg)	0.25
Fiber(g)	4.0	Riboflavin (mg)	0.01
Carbohydrates (g)	15.8	Vitamin C (mg)	9
Minerals (g)	0.8	Nicotinic acid (mg)	0.8
Magnesium (mg)	34	Oxalic acid (mg)	14
Phosphorus (mg)	139		
Sodium(mg)	7.8		
Calcium (mg)	20		
Vitamin A (IU)	139		
Copper(mg)	0.23		
Potassium(mg)	79		

3. Suitable Varieties

Early Smooth Seeded

Asauji, Lucknow Boniya, Alaska, Early Superb, Meteor,

Early Wrinkled Seeded

Arkel, Early Badger, Little Marvel, Kelvedon Wonder, Early December,

Wrinkled Seeded Main Season and Late Types

Bonneville, T-19, Lincolin, Khapar Kheda, NP 29, Thomas Laxton, Alderman, Kanwari,

Edible Podded Cultivar

Sylvia, Mattar Ageta 6, Aparna, Oregon 523, Taichung 12, Taichung 13.

4. Maturity Indices

Snow Peas are selected for size and maximal recovery of bright green, flat pods with minimal seed enlargement. Older and yellowing pods are avoided by careful hand-harvesting.

Sugar Snap Peas are selected in a similar manner but some degree of seed-pod filling is desirable. Larger seeds rapidly become starchy.

5. Quality Indices

Edible-pod peas should be uniformly bright green (light to deep green but not yellow-green), fully turgid, clean, and free from damage (Thrip injury, broken pods). The stem and calyxes should be green and there should be very few blossoms attached to the pods.

6. Postharvest Management

Harvesting

Harvesting green pods must be done at proper stage. The green pod pickings may be done during the morning or evening. The pea plants are very tender with soft stem and therefore picking should be done very gently. A small gerk damages the vines thereby injuring the plants. Moreover repeated picking should not be done. Only 2 pickings in early lines and 3 in mid season should be done. Adequate packaging may be done in gunny bags, baskets lined with jute cloth, bamboo baskets, corrugated fibre board boxes and plastic containers.

Storage

(*a*) Optimum Temperature

0°C; 95-98 per cent R.H.

Edible-pod peas are highly perishable and will not maintain good quality for more than 2 weeks. Wilting, yellowing of pods, loss of tenderness, development of starchiness and decay are likely to increase following storage beyond 14 days; defects occur faster at common distribution conditions of 5 to10°C.

(*b*) Rates of Respiration

Temperature °C	*ml CO_2/kg hr*
0	15–24
5	27–38
10	34–59
15	89–101
20	123–180

Respiration rates for edible-pod peas are an approximation based on values for unshelled garden peas; actual values remain to be determined.

To calculate heat production, multiply ml CO_2/kg·hr by 122 to get kcal/metric ton/day.

(*c*) Rates of Ethylene Production

< 0.1μl/kg·hr at 20°C

(*d*) Responses to Ethylene

Peas are moderately sensitive to exogenous ethylene. Accelerated yellowing and decay will result from extended exposure to low levels of ethylene during distribution and short-term storage. The calyx is more sensitive to ethylene than the pod.

(*e*) Responses to Controlled Atmosphere (CA)

Reports vary widely in the benefit of CA for Sugar and Snap Peas. Atmospheres of 2 to 3 per cent O_2 and 2 to 3 per cent CO_2 are considered by UC Research to offer the best, but moderate, benefit to peas beyond that of rapid cooling and proper storage. Low O_2 may promote off-flavors and off-odors. Other studies report that 5 to 7 per cent CO_2 extends pod quality at 0°C.

7. Postharvest Disorders, Diseases and their Control

Physiological Injury Disorders

Freezing

Freezing injury will be initiated at–0.6°C. Freezing injury results in watersoaking typically followed by rapid decay by soft-rot bacteria.

Premature Senescence

(Yellowing of pod, browning of calyx, loss of tenderness) will develop rapidly at temperatures 7.5°C due to the high rate of respiration.

Physical Injury

Harvesting and handling should be done with care to prevent damage to the pods and attached calyx.

Pathological Disorders

A variety of fungal pod-spotting and decay pathogens affect edible-pod peas. Common diseases include Chocolate Spot and Grey mold

Frost Damage

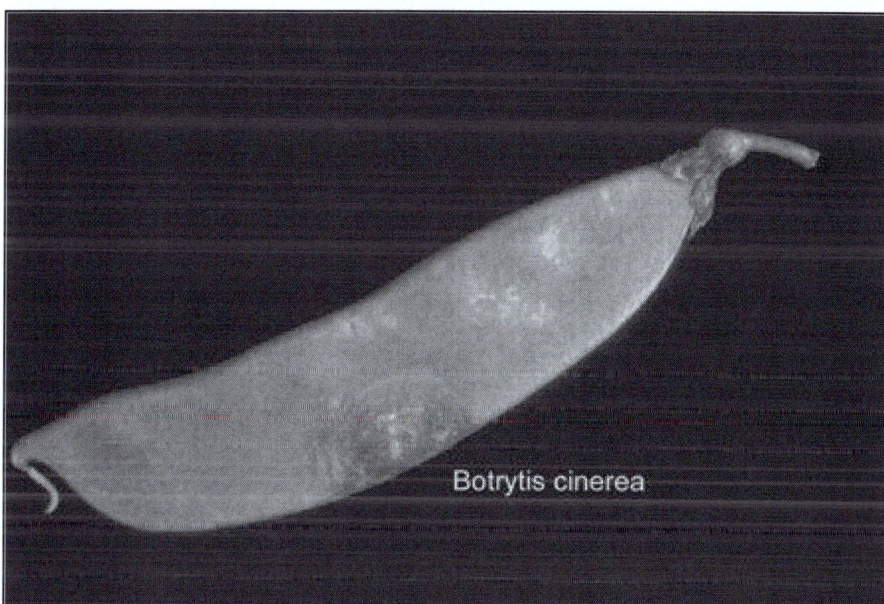

Grey Mold

(*Botrytis cinerea*), Watery Soft Rot (*Sclerotinia sclerotiorum*), Rhizopus Rot, and Ascochyta Pod Spot. Bacterial Soft Rot is common following rough handling or freezing injury. Surface decay can occur readily, on weak calyxes(brown at harvest) and on blossom remains.

Important to Note

Package-icing and top-icing loads may be used for Snow Peas but is typically deterimental to Snap Peas because surface moisture promotes decay. Improper CA/MA conditions in ready -to-cook vegetable medleys often leads to off-flavors and fungal decay (typically Botrytis grey mold) at the blossom-end of the pod.

37

Pumpkin

Botanical Name: ***Cucurbita moschata***

Family: Cucurbitaceae

1. Introduction

Pumpkin occupies a prominent place among vegetables owing to its high productivity, nutritive value, good storability, long period of availability, better transport qualities and extensive cultivation in subtropical and tropical parts of the world. It is used both in immature and mature stages as a vegetable. It is also consumed as processed and stock feed. The flesh is delicious when fried, stewed, boiled or baked. The fruits are sweetish when fully mature and can be sued in preparing sweets, candy or fermented into beverages. Yellow or orange-fleshed pumpkins are rich in carotene. Its young leaves, tender stem and flowers are also cooked and consumed. In India, it is grown mainly in Assam, West Bengal, Tamil Nadu, Karnataka, Madhya Pradesh, Uttar Pradesh, Orissa and Bihar.

2. Composition (Per 100g of Edible Portion)

Constituent	*Composition*	*Constituent*	*Composition*
Moisture (%)	95.8	Calcium (mg)	11
Protein(g)	0.2g	Phosphorus (mg)	12
Fat(g)	0.2 g	Iron(mg)	7.9
Minerals(g)	0.3 g	Thiamine(mg)	0.02
Fiber (g)	0.2 g	Riboflavin (mg)	0.04
Carbohydrates(g)	3.3 g	Niacin(mg)	0.I
Energy K Cal 16	Vitamin C(mg)	1	

3. Suitable Varieties

Ambili, Arka chandan, CO1, CO2, Pusa Hybrid 1, Pusa Vikas, Pusa Vishwas.

4. Maturity Indices

Corking of the stem and subtle changes in rind color (bright green to dull green) are the main external indications of maturity. Immature fruit have a fleshy stem, maturing fruit will have some stem corking, and well mature fruit will have a wellcorked stem. Internal color should be intense and typical of the cultivar. The concentrations of the yellow and orange carotenoids generally increase only slightly during storage. Maturity at harvest is the major determinant of internal color. Immature fruit will be of inferior eating quality because they contain less stored carbohydrates. Immature fruit will have more decay and weight loss during storage than mature fruits.

5. Quality Indices

Pumpkin and winter squash should be full sized and well formed with the stem intact. They should be well matured with good rind development typical of the cultivar. Internal quality attributes are high color due to a high carotenoid content, and high dry weight and sugar and starch contents.

6. Postharvest Management

Harvesting

Pumpkins are harvested manually by cutting it from the vine. The harvested fruits are cleaned. The deformed and wounded fruits should be discarded. The fruits are graded according to shape, colour and maturity.

Storage

(*a*) Optimum Temperature

12.5-15°C

Pumpkins are very chilling sensitive when stored below 10°C. Depending on the cultivar a storage life of 2 to 6 months can be expected at 12.5-15°C. Recent research at Oregon State for green rind squashes, storing at 15°C may cause degreening, undesirable yellowing, and texture loss. The green rind squashes can be stored at 10-12°C to prevent degreening, although some chilling injury may occur at the lower temperature. High storage temperature (>15°C) will result in excessive weight loss, color loss and poor eating quality.

(*b*) Optimum Relative Humidity

50-70 per cent with 60 per cent usually considered optimum. Moderate relative humidity with good ventilation is essential for optimum storage. High humidity will promote decay. Although 50-70 per cent RH will reduce decay during storage, significant weight loss will occur. For example, mature Kabocha squash lose 1.0 and 1.5 per cent of their fresh weight per week of storage at 12.5°C and 20°C, respectively.

(*c*) Rates of Respiration/CO_2 Production

30-60 ml CO_2/kg·hr at 25°C

To calculate heat production, multiply ml CO_2/kg·hr by 122 to get kcal/metric ton/day.

(*d*) Rates of Ethylene Production

<0.5µL C_2H_4/kg·hr at 20°C. If the fruits are chilled, ethylene production rates can be 3-5 times higher.

(*e*) Responses to Ethylene

Exposure to ethylene will degreen squash with green rinds. Ethylene will also cause abscission of the stem, especially in less mature fruit.

(f) Responses to Controlled Atmosphere (CA)

Atmospheres containing 7 per cent CO_2 can be beneficial by reducing loss of green color.

Yellow squash, however, appear not to be benefited by 5 or 10 per cent CO_2 atmospheres.

Lowering the O_2 concentration does not appear to provide any benefit.

7. Postharvest Disorders, Diseases and their Control

Physiological Disorders

Chilling Injury

Caused if pumpkins and squashes are stored below 10-12.5°C. Symptoms of chilling injury are sunken pits on the surface and high levels of decay once fruits are removed from storage. Storing fruit 1 month at 5°C is sufficient to cause chilling injury symptoms. Depending on the cultivar, storage for several months at 10°C may cause some chilling injury.

Freezing Injury

Can occur at temperatures below -0.8°C.

Pathological Disorders

Several fungi are associated with decay during storage of pumpkins and winter squashes. *Fusarium, Pythium* and anthracnose (*Colletotrichum*)

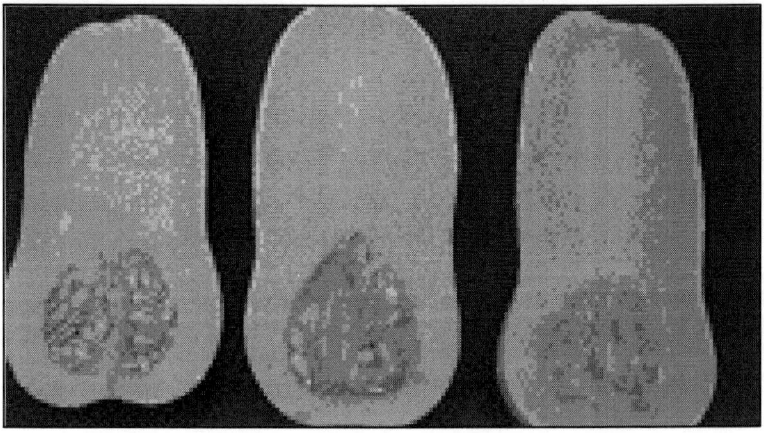

	1	2	3
L	74.4	70.5	66.9
Chroma	59.6	63.1	66.2
Hue	78.0	74.8	68.4

Butternut Squash Internal Color Scale

and gummy stem blight or black rot (*Mycosphaerella*) are common fungi. *Alternaria* rot will develop on chill-damaged winter squashes. Fruit that are overmature at harvest (>2 weeks beyond optimal harvest date) will tend to have more storage decay.

Important to Note

Curing

The fruits may have tender rinds when freshly harvested. Curing in the field (with protection from the sun by placing under the leaves) before handling and stacking into bins or wagons will help to harden or cure the rind. The recommended storage conditions also favor curing or hardening of the rind.

38

Radish

Botanical Name: ***Raphanus sativus*** **L.**

Family: Crucifereae

1. Introduction

Radish is a popular vegetable in both tropical and temperate regions. It is cultivated under glass for early production but large-scale production in the field is more common. Radish is one of the most ancient vegetables.

3. Composition (Per 100g Edible Portion)

Constituent	Composition			
	Radish Roots			
	White Cultivar	Pink Cultivar	Radish Tops	Radish Fruits
Moisture (%)	94.4	90.8	90.3	90.5
Protein (g)	0.7	0.6	2.7	2.3
Fat (g)	0.1	0.3	0.6	0.3
Fiber (g)	0.8	0.6	0.9	1. 4
Carbohydrates(g)	3.4	6.8	3.4	4.7
Minerals (g)	0.6	0.9	2.1	0.8
Calcium (mg)	50	50	310	80
Phosphorus (mg)	22	20	60	100
Iron (mg)	0.4	0.5	16.1	2.8
Vitamin A	5	5	1866.	50
Thiamine(mg)	0.06	0.06	0.03	0.07
Riboflavin (mg)	0.02	0.02	0.16	0.05
Nictonic acid (mg)	0.5	0.4	0.3	0.2
Vitamin C(mg)	15	17	103	69

2. Suitable Varieties

Pusa Desi, Pusa Himani, Pusa Chetki, Pusa Reshmi, Japanese White, Punjab Safed, Kalyani White, Chinese pink, IKR-1-1, Rapid red white tipped

3. Maturity Indices

Radish is a diversely formed root vegetable and has many uses worldwide. Red and icicle radish are most common but Asian "daikon" types are increasing in popularity outside of countries such as Korea, Japan, Taiwan and China. The number of days post-seeding or emergence, which may vary from 30 to 70 days, depending on type, typically determines maturity. A minimum size standard for common red radish is 5/8 inch (1.6cm) equatorial diameter. Current crop management practices stress rapid growth to ensure a mild flavor and crisp texture. Fertilization and irrigation management, or environmental conditions that slow growth may result in a woody texture and high pungency. Over-mature radish tends to be pithy (vacuolated) or spongy in texture and may develop harsh flavors, for most palates.

5. Quality Indices

Roots of Bunched or Topped Common Red Radish should, ideally, be of uniform and similar shape for the variety, well formed, smooth, firm but of tender texture, and free of growth or harvest damage, and free of decay, disease or insects. Bunched radish tops should be fresh in appearance, turgid, and free of freeze injury or other serious injury, seed stalk, yellowing or other discoloration, disease, decay, or insects.

6. Postharvest Management

Harvesting

It should be harvested when its roots are still tender. They are pulled out vertically with least breakage. In India radish is manually harvested. The delayed harvesting results in pithness and tastes bitter hence become unfit for market. The harvested roots along with tops are properly washed graded and tied in bundles. These bundles are loose packed in baskets and transported in the market. For washing, machines developed at PAU, Ludhiana can be effectively used with assured results.

Storage

(*a*) Optimum Temperature

Optimum temperature for its storage is 0 ± 1°C. Rapid cooling is essential to achieve the full storage potential of both bunched and topped roots. Radish is often top-iced to maintain temperature and contribute moisture for retaining a crisp texture. Under these conditions common red radish may be expected to maintain acceptable quality for 7 to 14 days with tops and 21 to 28 days if topped. Daikon-type radish may last from 3 to 4 months at these same conditions.

(*b*) Optimum Relative Humidity

95–100 per cent

(*c*) Rates of Respiration/CO_2 Production

Common Red Radish

Temperature	0°C	5°C	10°C	20°C
ml CO_2/kg hr				
Bunched	6–7	8–9	14–16	58–62
Topped	2–4	3–5	6–7	19–26

To calculate heat production multiply ml CO_2/kg hr by 122 to get kcal/metric ton/day.

(*d*) Rates of Ethylene Production

Very low; <0.1 mL/kg·hr at 20°C

(*e*) Responses to Ethylene

Not Sensitive

Bunched tops may exhibit yellowing with prolonged storage and ethylene exposure.

(*f*) Responses to Controlled Atmospheres(CA)

Atmospheres of 1 to 2 per cent O_2 and 2 to 3 per cent CO_2 are slightly beneficial in maintaining quality of topped radish when storage temperatures are 5 to 7°C. CA helps to retard the re-growth of shoots and rootlets in "topped and tailed" roots. Even short exposure to temperatures above 7°C will result in the development of off-flavors, browning, and soft-rot.

7. Postharvest Disorders, Diseases and their Control

Physiological Disorders

Freeze Injury

As radish is, ideally, stored and transported just above the freezing point (-1.0°C), freeze injury is not uncommon. Shoots become water-soaked, wilted, and turn black. Roots appear water-soaked and glassy, often only at the outer layers if the freezing temperature is not too low. Roots become soft quickly on warming and pigmented roots may "bleed" (lose pigment).

Pathological Disorders

Bacterial Black Spot

(*Xanthomonas campestris* cv. *vesicatoria*) is a problem in some production locations and will develop in postharvest storage at warmer than optimum temperatures. Refrigeration is the primary control but washing roots in chlorinated water is reported to significantly control this disease.

Prompt cooling, chlorination, and refrigeration are also effective in controlling Bacterial Soft Rot (*Erwinia carotovora* subsp *carotovora*).

Rhizotonia spp. lesions may develop in storage at warmer than optimal temperatures but is more effectively controlled in the field. *Botrytis* (Grey Mold) and *Sclerotinia* (Watery Soft Rot) can develop, especially around harvest wounds, even at temperatures below 5°C.

39

Spinach (Palak)

Botanical Name: ***Beta volgaris*** **var.** ***Bengalensis***

Family: Chenopodiaceae

1. Introduction

Palak is one of the most common leafy vegetables of tropical and subtropical region and is grown widely in India. The popular palak growing States are Uttar Pradesh, West Bengal, Punjab, Haryana, Delhi, Madhya Pradesh, Bihar, Maharashtra and Gujarat. However, this crop is not so popular in southern states.

2. Composition (Per 100g Edible Portion)

Constituent	*Composition*
Moisture (%)	86.4
Fat (g)	0.8 g
Fiber (g)	0.7 g
Protein (g)	3.4 g

Constituent	*Composition*
Minerals (g)	2.2 g
Carbohydrates(g)	6.5 g
Phosphorus (mg)	30 mg
Riboflavin (mg)	0.56 mg
Calcium (mg)	380 mg
Iron (mg)	16.2 mg
Thiamine(mg)	0.26 mg
Vitamin A (IU)	9770
Nicotinic acid (mg)	3.3 mg
Vitamin C (mg)	70 mg

3. Suitable Varieties

All Green, Pusa Palak, Pusa Jyoti, Pusa Hart, Jobner Green, HS-23 Palak No. 51-16, Banerjees Giant.

4. Maturity Indices

Spinach is selected for size and maximal recovery of clean leaves that are mid-maturity to young. Older and yellowing leaves are avoided when making the harvest cut. Generally 3-4 weeks of re-growth are required before a second harvest will yield adequate volume.

5. Quality Indices

Spinach, whether bunched or as leaves, should be uniformly green (generally not yellow-green), fully turgid, fairly clean, and free from serious damage. For bunched spinach, roots should be trimmed short to grade standards and petioles should be predominantly shorter than the leaf blade.

6. Postharvest Management

Harvesting

Harvesting is done in the evening because leaves become crisp due to dew and easily break in the morning. First cutting can be taken about 4 weeks after sowing. It is possible to get 4-6 cuttings/season at 15-20 days interval depending upon variety and season. After harvesting damaged and diseased leaves are removed and bundled in to bunches of about 10-20 leaves for convenience in handling and marketing.

Before storage the produce may be thoroughly washed. For washing machines developed at PAU, Ludhiana are found very effective.

Storage

(*a*) Optimum Temperature

0°C; 95-98 per cent R.H.

Spinach is highly perishable and will not maintain good quality for more than 2 weeks. Wilting, yellowing of leaves, and decay are likely to increase following storage beyond 10-14 days; faster at common distribution conditions of 5 to10°C.

(*b*) Rates of Respiration/CO_2 Production

Temperature °C	*ml CO_2/kg hr*
0	9-11
5	17-29
10	41-69
15	67-111
20	86-143

To calculate heat production, multiply ml CO_2/kg hr by 440 to get BTU/ton/day or by 122 to get kcal/metric ton/day.

(*c*) Rates of Ethylene Production

< 0.1µl/kg·hr at 20°C

(*d*) Responses to Ethylene

Spinach is highly sensitive to exogenous ethylene. Accelerated yellowing will result from low levels of ethylene during distribution and short-term storage. Do not mix loads such as apples, melons and tomatoes with spinach.

(*e*) Responses to Controlled Atmosphere (CA)

Atmospheres of 7-10 per cent O_2 and 5-10 per cent CO_2 offer moderate benefit to spinach by delaying yellowing. Spinach is tolerant to higher CO_2 concentration but no increase in benefits has been observed. Package film for prewashed spinach leaves is selected to maintain 1-3 per cent O_2 and 8-10 per cent CO_2.

7. Postharvest Disorders, Diseases and their Control

Physiological Disorders

Freezing Injury

Freezing injury will be initiated at–0.3°C. Freezing injury results in watersoaking typically followed by rapid decay by soft-rot bacteria.

Yellowing

Spinach is highly sensitive to exogenous ethylene.

Physical Injury

Harvesting and handling should be done with care to prevent damage to the petioles and leaves. Bunching ties should not be too tight as crushed or spilt petioles may lead to rapid decay.

Pathological Disorders

Bacterial Soft-Rot (primarily *Erwinia* and *Pseudomonus*) is a common problem. Decay is usually associated with damaged leaves and stems.

Important to Note

Package-icing and top-icing loads may be used. Frequent light misting may be done in displays to delay wilting of bunched spinach.

40

Sweet Potato

Botanical Name: ***Ipomoea batatus*** **(L)**

Family: Convolvulaceae

1. Introduction

The sweet potato is an important starchy food crop thoughout the tropical and subtropical countries. It is popularly called 'Sakarkand' in India. The speciality of this crop to adapt under different climatic zones. The high orange-colored flesh contain much higher levels of carotenoids than less pigmented types. Sweet potato flavor is largely based on starch and sugar concentrations, and these are affected by cultivars and storage conditions. The leading States in terms of area and production are Bihar, Orissa and Uttar Pradesh. The other major sweet potato growing states in India are Maharashtra, Karnataka and Madhya Pradesh. Sweet potato is grown in almost all the states of India except Jammu and Kashmir.

2. Composition (Per 100g of Edible Portion)

Constituent	Composition	
	Tuber (per cent)	Vine (per cent)
Moisture (%)	70	80.7
Carbohydrates (g)	27	9.7
Protein (g)	1.5-2.0	4.2
Fat (g)	0.2	0.8
Sugar (g)	3-6	2.4
Calcium (mg)	46	360
Phosphorus (mg)	49	60
Iron (mg)	0.8	10
Magnesium (mg)	24	–
Potassium (mg)	373	–
Sodium (mg)	13	–
Chlorine (mg)	85	–
Sulphur (mg)	26	–
Thiamine (mg)	0.08	0.07
Riboflavin (mg)	0.04	0.24
Niacin (mg)	0.70	1.70
Vitamin C (mg)	24.0	27
Carotene (mg)	0-18	4
Energy (Kcal)	120	63

3. Suitable Cultivars

H-620, Cross-4, R.S.-5, Pusa Safed, V-35, Kalmegh, H-635, S-30, H-41, H-42, H-268.

4. Maturity Indices

Sweet potato matures in 100 to 125 days after planting. Harvesting sweet potato 120 days after planting is normally recommended. Delay in harvesting invites attack of sweet potato weevil. Maturity is indicated when the leaves turn yellow and begin to fall. By cutting tubers and verifying that latex dries up without turning black indicates its maturity.

5. Quality Indices

Good quality sweet potatoes should be smooth and firm, with uniform shape and size, be free from mechanical damage, and have a uniform peel color typical of the variety. Grades are based on degree of freedom from defects (dirt, roots, cuts, bruises, growth cracks, decay, insects, and diseases), but also size and weight categories.

6. Postharvest Management

Harvesting

For easy harvesting light irrigation is given 2-3 days before digging of the tubers. Care should be taken to avoid injuries and bruises to the tubers. For marketing of the fresh tubes cleaning and grading should be done to get more prices. After harvest tubers are spread in partial shade for 5-6 days for healing and curing.

Storage

(*a*) Optimum Temperature

The recommended conditions for commercial storage are to keep roots cool and dry. Sweet potato roots are chilling sensitive and should be stored between 12.5°C and 15°C with high relative humidity (>90 per cent). A storage life of 6-10 months can be expected under these conditions, although sprouting may begin to occur after about 6 months depending on cultivar. Temperatures above 15°C lead to more rapid sprouting and weight loss. Careful handling during harvesting will minimize mechanical damage to the skin and reduce decay incidence during storage. Roots are not washed before storing in bins or crates, but only after removal for selection and packing for marketing. Sweet potato roots are commonly stored in evaporatively cooled rooms, supplemented by mechanical refrigeration late in the storage period when warm ambient temperatures occur.

(*b*) Optimum Relative Humidity

> 95 per cent for long-term storage; 70-90 per cent for short-term handling for marketing

(*c*) Rates of Respiration/CO_2 Production

Temperature	10°C	15°C	25°C
ml CO_2/kg hr			
Cured	7	10–12	–
Noncured	–	15	27–35

To calculate heat production multiply mL CO_2/kg hr by 122 to get kcal/metric ton/day.

(*d*) Rates of Ethylene Production and Responses to Ethylene

Sweet potato roots produce very low amounts of ethylene (~0.1 µL/kg·hr), although much higher rates can occur after chilling, wounding and decay development. Exposure to ethylene (1 to 10 ppm) increases respiration rates and phenolic metabolism and adversely affects flavor and color of cooked roots.

(*e*) Responses to Controlled Atmospheres (CA)

There is no commercial use of controlled atmospheres for sweet potato storage. Respiration rates of roots are reduced as oxygen is lowered from 21 to 3 per cent. Oxygen concentrations below 3 per cent may results in increased respiration rates due to fermentative metabolism. Response of roots to increase carbon dioxide levels is not known.

7. Postharvest Disorders, Diseases and their Control

Physiological Disorders

Chilling Injury

Sweet potato roots are very sensitive to chilling injury at temperatures of 12.5°C or below. Symptoms of chilling injury include fungal decay, internal pulp browning, and root shriveling. Chilled roots that have been cooked can have "hardcore" defect and a darker color than non-chilled roots.

Pathological Disorders

Chilling and mechanical injury predispose sweet potatoes to decay, especially *Rhizopus* soft rot. Postharvest fungicides may be applied to

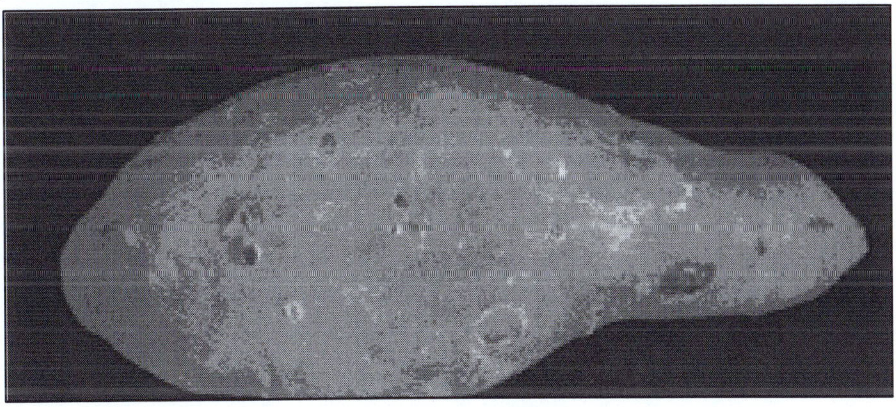

Botryodiplodia Rot

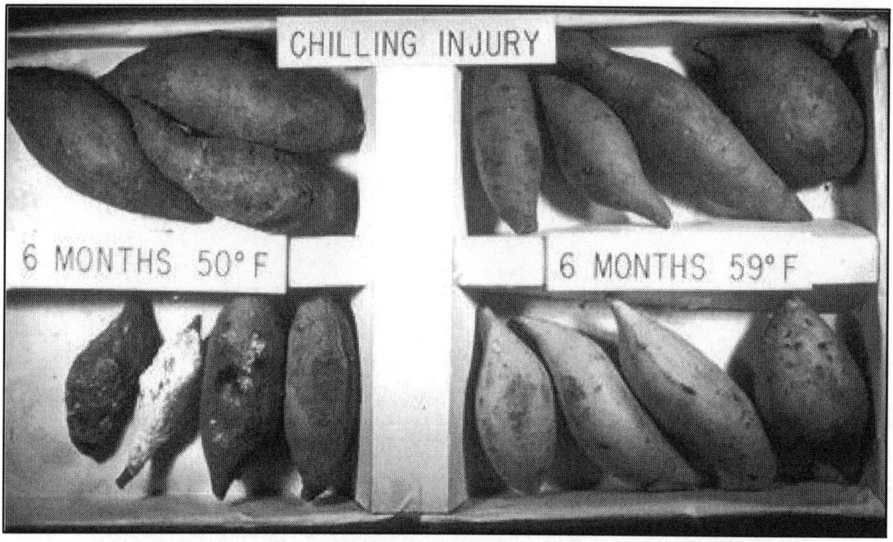

Chilling Injury

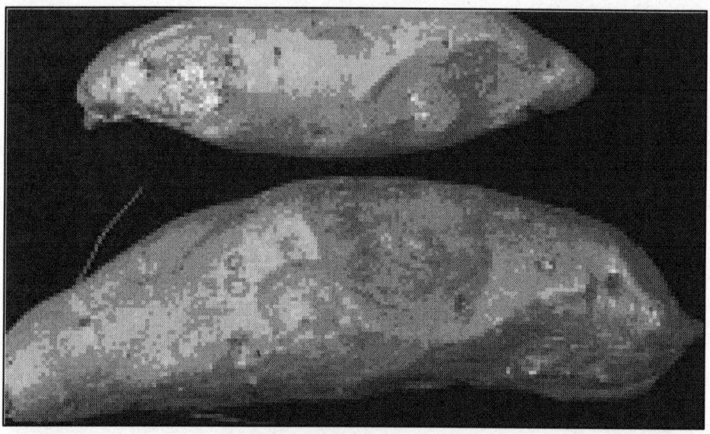

Fusarium Rot

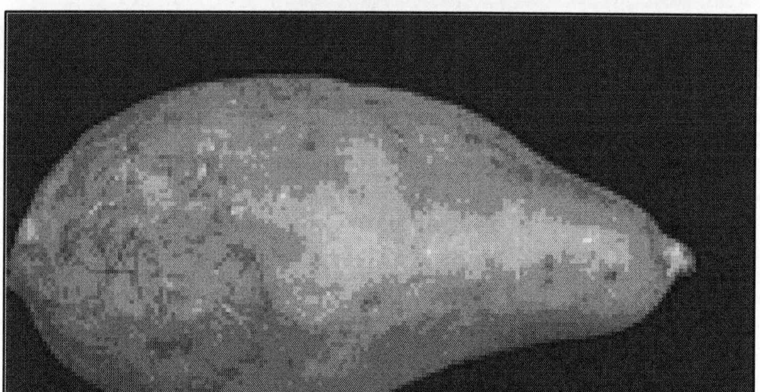

Rhizopus Rot

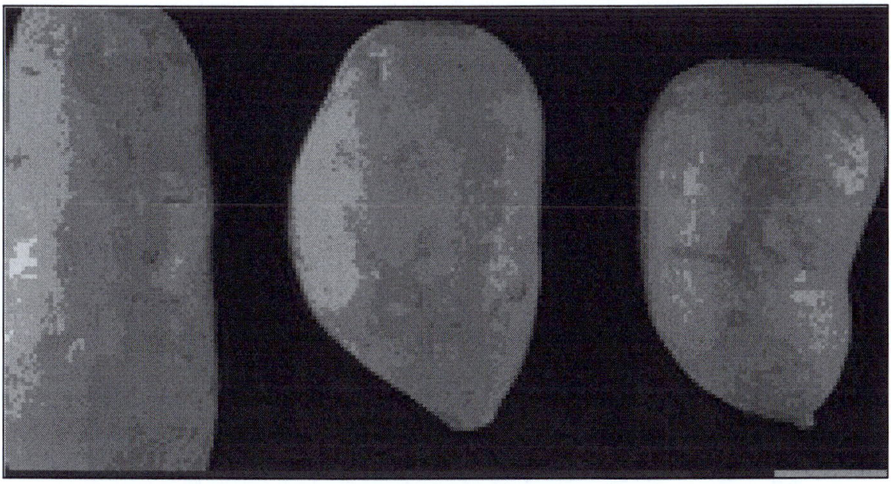

Surface Abrasion

reduce the risk of Rhizopus after handling for marketing. There are numerous other decay-causing fungi including black rot (*Ceratocystis*) and Fusarium rot. Seed piece treatment and postharvest curing are the main control measures for these organisms. In warm wet production conditions, bacterial rots can also cause postharvest losses.

Important to Note

Curing

The periderm of sweet potato roots is easily damaged during harvest and handling, and this leads to an unsightly appearance, high rates of water loss, and increased susceptibility to decay. The process of curing the damaged skin or "wound healing" can be achieved by holding roots at 25-32°C under high relative humidity (>90 to 100 per cent) for several days to 1 week. The conditions for curing sweet potatoes are similar to those used for other tropical root and tuber crops. Growers often load bins of warm roots into storage rooms and do not turn on the fans for evaporative cooling until after about 1 week. This interval before cooling provides the warm humid conditions necessary for curing wounds.

41

Tomato

Botanical Name: ***Lycopersicon esculentum***

Family: Solanaceae

1. Introduction

Tomato is one of the most popular and widely grown vegetables in the world ranking second in importance to potato in many countries. The fruits are eaten raw or cooked. Large quantities of tomato are used to produce soup, juice, ketchup, purre, paste and powder. Tomato is popular also because it contains vitamin C and lycopene, which adds variety of colours and flavours to the foods and also reduces risk of several important chronic diseases, including coronary heart disease and a number of types of cancer associated with nutritional traditions, such as breast, colon, and prostate cancer. Green tomatoes are also used for pickles and preserves. Its many forms are adapted to wide range of soils and climates and its culture extends from the tropics to a few degrees within the Arctic Circle. It has many other uses, as tomato seed contains 24 per cent oil and this is extracted from the pulp and residues in canning industry. The semidrying oil is used as salad oil and in the manufacture of margarine.

2. Composition (Per 100g Edible Portion)

Constituent	*Composition*	*Constituent*	*Composition*
Moisture (%)	93.1	Riboflavine (mg)	0.01
Protein (g)	1.9	Nicotinic acid (mg)	0.4
Fat (g)	0.1	Vitamin C (mg)	31
Minerals(g)	0.6	Calcium (mg)	20
Fiber(g)	0.7	Magnesium (mg)	15
Carbohydrates(g)	3.6	Oxalic acid (mg)	2
Sodium (mg)	45.8	Phosphorus (mg)	36
Potassium(mg)	114	Iron (mg)	1.8
Copper (mg)	0.19		
Sulphur (mg)	24		
Chlorine (mg)	38		
Vitamin A (IU)	320		
Thiamine(mg)	0.07		

3. Suitable Varieties

For Plains: Pusa Ruby, Pusa Early Dwarf, Pusa 120, HS 102, Sweet 72, S-12, CO-1,

For Hills: Sioux, Best of all, Marigold

For Processing: Roma, Punjab Chuara, S-152,

Hybrids: Rashmi, Rupali, IAHS-88.2, Naveen

4. Maturity Indices

Standard Tomatoes

Minimum harvest maturity (Mature Green 2) is defined by internal fruit structure indices. Seeds are fully developed and are not cut upon slicing the fruit. Gel formation is advanced in at least one locule and jelly like material is forming in other locules.

Off-vine ripening is severely affected if fruits are harvested at the MG2 stage. Minimum harvest maturity is better defined as equivalent to ripeness class Pink

5. Quality Indices

Standard tomato quality is primarily based on uniform shape and freedom from growth or handling defects. Size is not a factor of grade quality but may strongly influence commercial quality expectations.

Shape

Well formed for type (round, globe, flattened globe, roma)

Color

Uniform color (orange-red to deep red; light yellow). No green shoulders.

Appearance

Smooth and small blossom-end scar and stem-end scar. Absence of growth cracks, catfacing, zippering, sunscald, insect injury, and mechanical injury or bruises.

MG1 MG2 MG3 MG4 Breaker

Maturity stages of tomato

MG: Mature-green

Tomato Maturity Stages

Firmness

Yields to firm hand pressure. Not soft and easily deformed due to an overripe condition.

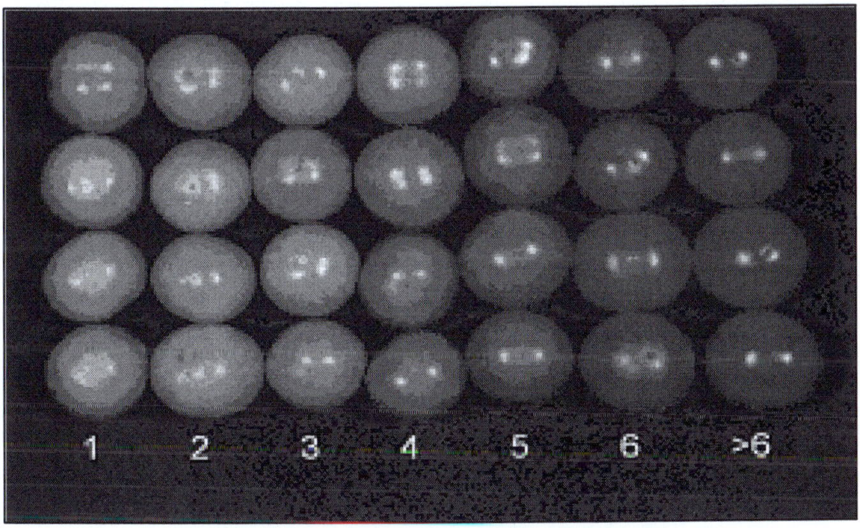

Cherry Tomato Maturity and Ripeness

Optimum Temperature

Mature Green: 12.5–15°C

Light Red: 10–12.5°C

Firm-ripe: 7–10°C for 3-5 days

Mature-green tomatoes can be stored up to 14 days prior to ripening at 12.5°C without significant reduction of sensory quality and color development. Decay is likely to increase following storage beyond two weeks, at this temperature. Typically 8-10 days of shelflife is attainable within the optimum temperature range after reaching the Firm-ripe stage. Short-term storage or transit temperatures below this range are used by some in the trade but will result in chilling injury after several days. Extended storage with controlled atmosphere has been demonstrated.

7. Postharvest Management

Harvesting

Tomatoes are harvested at a variety of stages of ripeness, from mature green to light pink, when they are easily separated from the vine by a half turn or twist. Commercially, most of the fresh market tomatoes are harvested by hand.

Precooling

After harvesting in order to remove the field heat tomatoes are cooled rapidly at 13°C, which in turn increases its shelf life.

Storage

(*a*) Ripening Temperatures

18° -21°C; 90-95 per cent R.H. for standard ripening 14° -16°C for slow ripening (*i.e.* in transit).

(*b*) Chilling Injury

Tomatoes are chilling sensitive at temperatures below 10°C if held for longer than 2 weeks or at 5°C for longer than 6-8 days. Consequences of chilling injury are failure to ripen and develop full color and flavor, irregular (blotchy) color development, premature softening, surface pitting, browning of seeds, and increased decay (especially Black mold caused by *Alternaria* spp.). Chilling injury is cumulative and may be initiated in the field prior to harvest.

(*c*) Optimum Relative Humidity

90-95 per cent; High relative humidity is essential to maximize postharvest quality and prevent water loss (desiccation). Extended periods of higher humidity or condensation may encourage the growth of stem-scar and surface molds.

(*d*) Rates of Respiration/CO_2 Production

Temperature	*ml CO_2/kg hr*	*Mature-green Ripening*
5°C	3–4[NR]	
10°C	6–9	7–8
15°C	8–14	12–15
20°C	14–20	12–22
25°C	18–26	15–26

To calculate heat production, multiply ml CO_2/kg hr by 122 to get kcal/metric ton/day.

NR: Not recommended for more than a few days due to chilling injury.

(*e*) Rates of Ethylene Production

1.2–1.5µl/kg·hr at 10°C

4.3–4.9µl/kg·hr at 20°C

(*f*) Responses to Ethylene

Tomatoes are sensitive to exogenous ethylene and exposure of mature-green fruit to ethylene will initiate ripening. Ripening tomatoes produce ethylene at a moderate rate and co-storage or shipment with sensitive commodities, such as lettuce and cucumbers, should be avoided.

(*g*) Ripening

Faster ripening results from higher temperatures between 12.5 -25°C; 90-95 per cent R.H.; 100 ppm ethylene. Good air circulation must be maintained to ensure temperature uniformity within the ripening room and to prevent the accumulation of CO_2. CO_2 (above 1 per cent) retards the action of ethylene in stimulating ripening.

The optimum ripening temperature to ensure sensory and nutritive quality is 20°C. Color development is optimal and retention of vitamin C content is highest at this ripening temperature. Tomatoes allowed to ripen off-the-vine above 25°C will develop a more yellow and less red color and will be softer.

Ethylene treatment typically extends for 24-72 hours. A second treatment period may follow repacking if immature green fruit were included in the harvest.

(*h*) Responses to Controlled Atmospheres (CA)

Controlled atmosphere storage or shipping offer a moderate level of benefit. Low O_2 levels (3-5 per cent) delay ripening and the development of surface and stem-scar molds without severely impacting sensory quality for most consumers. Storage times of up to 7 weeks have been reported for tomatoes using a combination of 4 per cent O_2, 2 per cent CO_2, and 5 per cent CO. More typically, 3 per cent O_2 and 0-3 per cent CO_2 are used to maintain acceptable quality for up to 6 weeks prior to ripening. Elevated CO_2 above 3-5 per cent is not tolerated by most cultivars and will cause injury. Low O_2(1 per cent) will cause off-flavors, objectionable odors, and other condition defects, such as internal browning.

7. Postharvest Disorders, Diseases and their Control

Physiological Disorders

Freezing Injury

Freezing injury will be initiated at -1°C, depending on the soluble

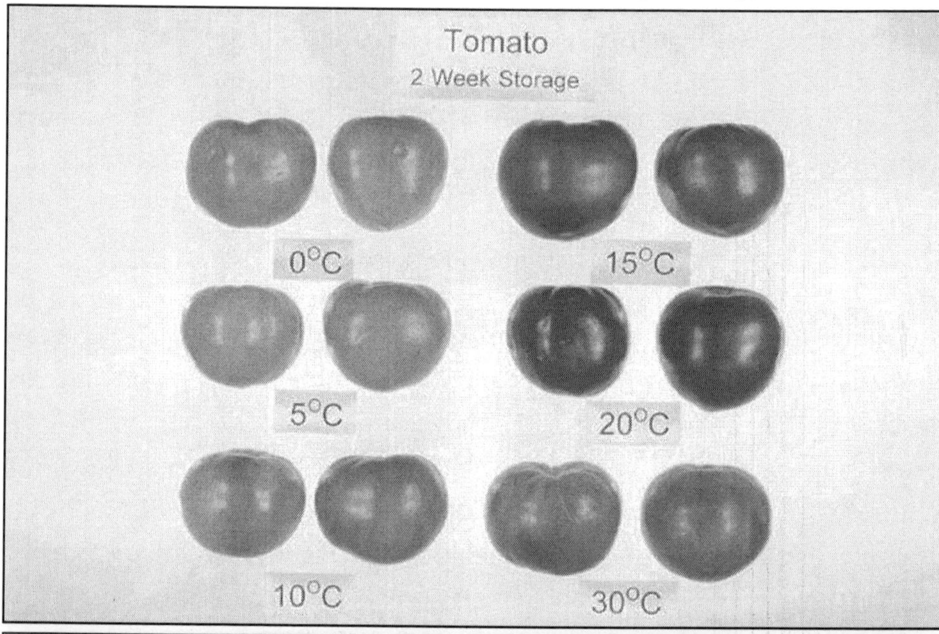

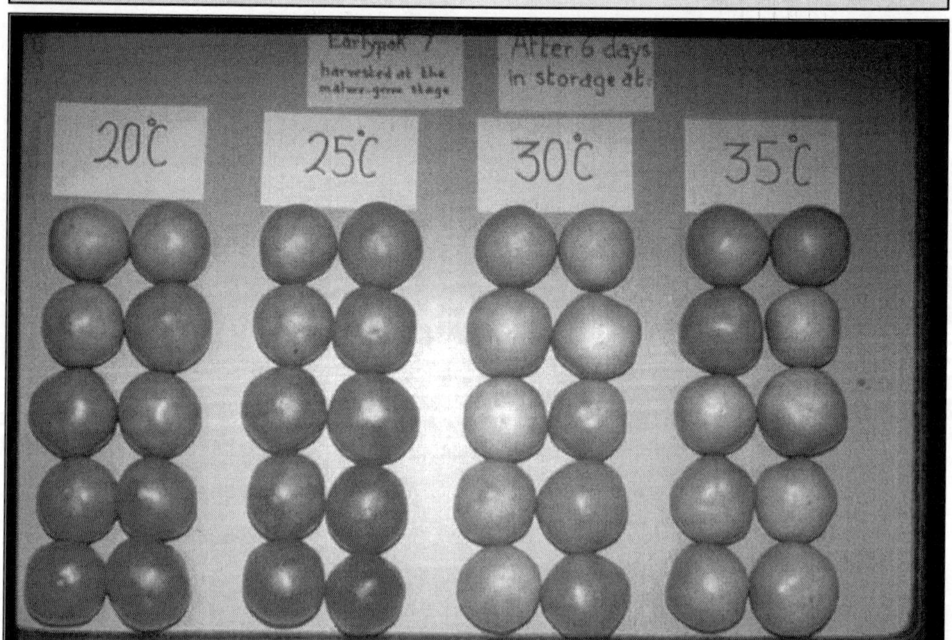

Effects of Temperature on Tomato

Effects of High Temperatures on Tomatoes

solids content. Symptoms of freezing injury include a watersoaked appearance, excessive softening, desiccated appearance of the locular gel.

Physical Disorders

Tomatoes are sensitive to many production and environment-genetic interaction disorders which may disorders be manifested during postharvest ripening or postharvest inspection. Fertilizer and irrigation management, weather conditions, insect feeding injury, asymptomatic virus infection, and unknown agents may interact to affect postharvest quality. Examples are Blossom-end Rot, Internal White Tissue, Rain Checking, Concentric and Radial Cracking, Puffiness, Persistent Green Shoulder, and Graywall. Several references with photographic keys to disorders are available.

Pathological Disorders

Diseases are an important source of postharvest loss depending on

Control Ethephon **Ethephon Effects on Tomato**

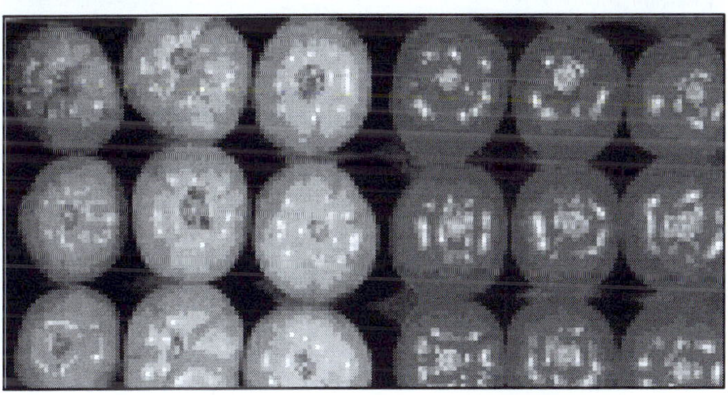

Chilled Control **Chilling Injury of Tomatoes**

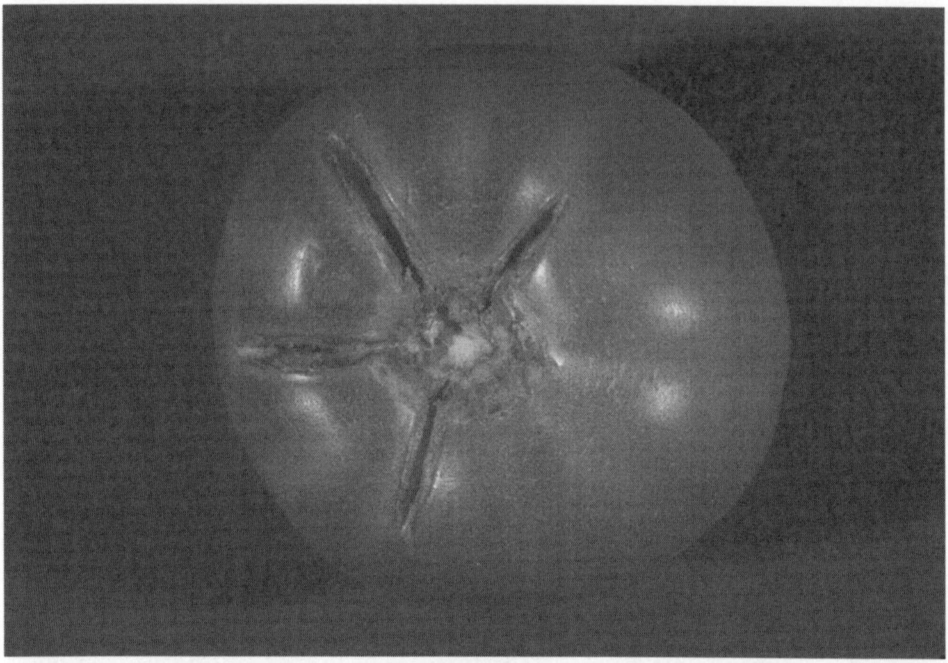

Tomato Cracking

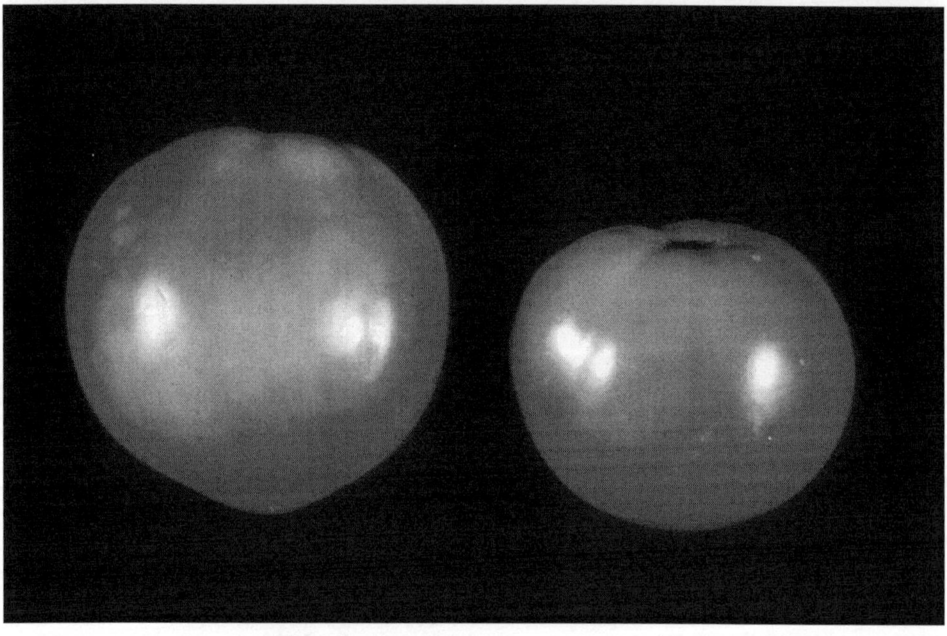

Tomato Solar Yellowing Damage

season, region and handling practices. Commonly, decay or surface lesions result from the fungal pathogens *Alternaria* (Black Mould Rot), *Botrytis*

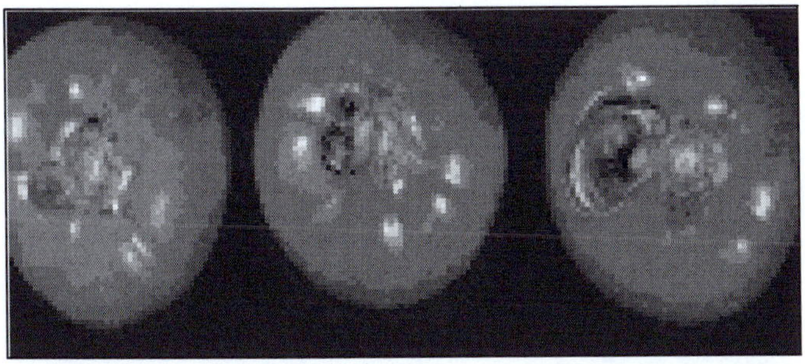

Tomato Alternaria from Chilling (After 10 days at 5°C)

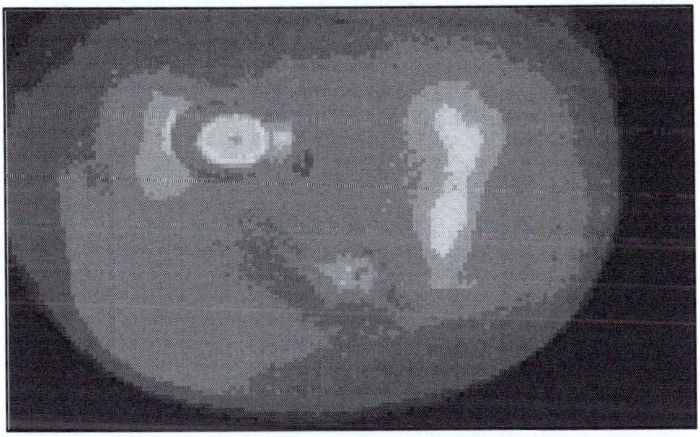

Tomato Athracnose

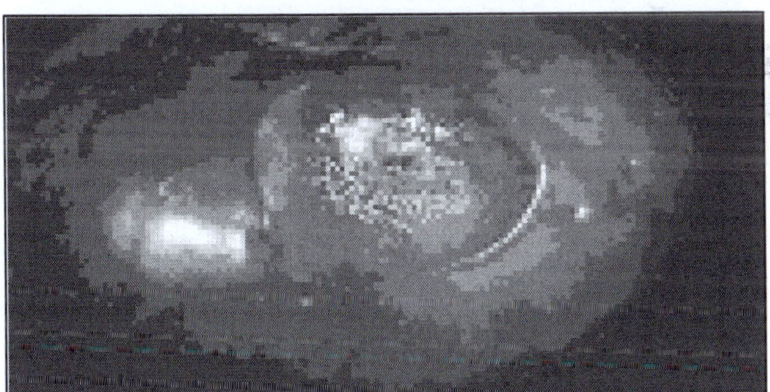

Tomato Fusarium

(Gray Mould Rot), *Geotrichum* (Sour Rot), and *Rhizopus* (Hairy Rot). Bacterial Soft Rot caused by Erwinia spp. can be a serious problem particularly if proper harvest and packing house sanitation is not used. Treatment with hot air or hot water immersion (55°C for 0.5–1.0 min.) has been effective in preventing surface mold but has not been used

Tomato Phytophthora

extensively for commercial treatments. CA can be effective in delaying fungal growth on the stem-end and fruit surface.

Greenhouse tomatoes marketed on-the-vine ("cluster tomatoes") are very susceptible to Botrytis Gray Mold, especially if film-wrapped in a tray.

Important to Note

Rapid cooling soon after harvest is essential for optimal postharvest keeping quality. The precooling endpoint is typically 12.5°C. Forced-air cooling is the most effective practice but room cooling is more common.

42

Potato

Botanical Name: *Solanum tuberosum*

Family: Solanaceae

1. Introduction

Potato is an important starch food crop in both sub-tropical and temperate regions. Even in tropical region it is widely grown during winter season. It occupies the largest area under any single crop in the world and it produces more food per unit area than cereals and that too in a

short time. In some respects, it excels cereals in nutritive value and palatability.

2. Composition

Constituent	Composition	Constituent	Composition
Moisture (%)	77.2	Vitamin C (mg)	25
Crude protein (g)	2.8	Thiamine (mg)	0.1
True protein (g)	1.6	Riboflavin (mg)	0.02
Starch (g)	16.3	Ascorbic acid (mg)	17
Total sugar(g)	0.6	Nicotinic acid (mg)	0.5
Reducing sugar(g)	0.3	Carotene (mg)	24
Energy (MJ)	0.3	Essential amino acids	
Crude fiber (g)	0.59	Arginine (g)	4.6
Crude fat(g)	0.14	Cystine (g)	0.8
Minerals (g)	0.9	Histidine (g)	1.5
Calcium (mg)	7.7	Leucine (g)	10.4
Copper m(g)	0.15	Lycine (g)	6.2
Iron (mg)	0.75	Methionine (g)	2.8
Magnesium (mg)	24.2	Phenylaanine (g)	4.4
Phosphorus (mg)	40.3	Threonine (g)	4.8
Potassium (mg)	568.0	Tryptophan (g)	1.6
Sodium (mg)	6.5		

3. Suitable Varieties

Kufri Sindhuri, K. Chandaramukhi, K. Jyoti, K. Muthu, K. Lauvkar, K. Deva, K. Badhsha, K. Bahar, K. Lalima, Kufari Swarna, K. Megha, K. Ashoka, K. Jwahar, K. Sutlej

4. Maturity Indices

Immature potatoes, generally harvested in spring or early summer, have a thin, poorly developed periderm, or skin. Irrigation and planting bed management, along with the option of vine-killing treatments, manage harvest "maturity". Harvest preparedness generally begins once tubers have reached a desirable size for the variety or market. Immature potatoes are easily bruised and "skinning" leads to shriveling or decays. They are very perishable relative to late-crop potatoes and are only stored for short periods. Curing potatoes for 8 days at 15°C and 95 per cent RH will allow extended storage up to 5 months at 4°C and 95 to 98 per cent RH, depending on variety. More typically, early-crop potatoes are harvested, cooled at 15°C, treated with a sprouting inhibitor, packed and shipped in a short period of time (*i.e.* 1 to 5 days).

5. Quality Indices

High quality traits, in commercial trade, include more than 70 to 80 per cent of tubers well shaped, brightness of color (esp. reds, yellows, and whites), uniformity, firmness, freedom from adhering soil, freedom from bruising (black spot or shatter-bruising), scuffing or skinning, growth cracks, sprouting, insect damage, *Rhizoctonia* Black Scurf, decay, greening, or other defects. Commercial standards in use are typically higher than USDA grade standards. Differentiation of quality for potatoes is very complex.

6. Postharvest Management

Harvesting

Potato tubers are harvested as soon as they mature. However, time of harvest can be adjusted to suit market prices and demand. Manual harvesting is the common practice. In large coverage two row mechanical diggers or mechanical harvesters are used. In two-row digger potatoes are usually dropped to the ground behind the digger and later on picked up by hands. They are placed in to containers, sacks or directly in to crates and loaded in to trucks and transported to the storage or packing shed. Losses of potato depend upon maturity stage and method of harvesting.

Storage

Before storage the potato tubers are thorough washed using vegetable and fruit washing machines developed at AICRP (PHT) centers.

(*a*) Optimum Storage Conditions

Intended Use	*Temperature*	*%RH*
Table	7°C	98
Frying	10 to 15°C	95
Chipping	15 to 20°C	95

At optimum conditions, potatoes should have good quality after storage of 3 to 5 weeks. Storing immature potatoes below 10-13°C for as few as 3 days may cause the accumulation of reducing sugars leading to excessive browning during frying/chipping. Storage for less than 3 weeks is recommended to maintain good visual and sensory quality of immature potatoes.

(*b*) Rates of Respiration/CO_2 Production

Temperature °C	*ml CO_2/kg hr**
5	6–8
10	7–11
15	7–16
20	9–23

* To calculate heat production, multiply ml CO_2/kg by 122 to get kcal/metric ton/day.

Note: Immature potato tubers, which are susceptible to bruising and skinning, can have high respiration rates. Cooler temperatures and/or increased air movement are effective methods to ameliorate.

(*c*) Rates of Ethylene Production

Very low;<0.1 µl/kg·hr at 20°C

Bruised, cut or otherwise damaged tubers have greatly increased ethylene production rates.

(*d*) Responses to Ethylene

Potato tubers are not very sensitive to external ethylene. Low levels of external ethylene have been shown to elevate respiration, especially in immature potatoes, and result in weight loss and mild shriveling. After aging for 2-3 months at temperatures above 5°C and in the absence of sprouting inhibitor application, low levels of ethylene may retard sprouting. High concentrations of external ethylene may induce sprouting.

(*e*) Responses to Controlled Atmosphere (CA)

Controlled or modified atmospheres offer little benefit to potato. Periderm development and wound healing is delayed at atmospheres below 5 per cent O_2. Injury from low O_2 Atmosphere (CA) (< 1.5 per cent) or elevated CO_2 (>10 per cent) will induce off-odors, off-flavors, internal discoloration, and increased decay.

7. Postharvest Disorders, Diseases and their Control

Physiological Disorders

Blackheart

Rare in early-crop potatoes due to typical marketing practices; In conditions of restricted airflow and high respiration, tubers held above 15°C (rapidly above 20°C) develop an internal brown discoloration which eventually becomes deep black. Insufficient oxygen reaches the interior of the tuber under these conditions.

Black Spot

Responsible for significant postharvest losses, particularly in response to over-fertilization with nitrogen, low soil potassium availability, irregular irrigation, and other pre-harvest practices. Non-pigmented compounds form in the vascular bundle tissue just under the skin during storage. Following severe bruising or cutting, the affected tuber tissue turns reddish, then blue becoming black in 24 to 72 hours. Severity increases with time. Varieties differ significantly in their susceptibility and symptom expression.

Chilling Injury

Storage at temperatures near 0°C for a few weeks may result in a mahogany discoloration of internal tissue in some varieties. Much longer periods of storage are generally required for chilling injury.

Greening

Exposure to bright light during postharvest handling, or longer periods (1 to 2 weeks) of low light intensity, can result in the development of chlorophyll in the potato tuber, anatomically a modified stem. Associated with greening, bitter and toxic glycoalkaloids, such as solanine, are formed. Solanine also forms in response to bruising, wounding (including fresh processing followed by storage), and during sprouting. Glycoalkaloids are heat stable and minimally impacted by cooking.

Internal Brown Spot

Internal dry, corky reddish-brown or black spots or sectors. Uneven water management and/or widely fluctuating temperatures induce this calcium uptake deficiency, usually early in tuber development. Uneven water availability may also result in Hollow Heart, a corky cavitation at the center of the tuber.

Physical Injury

Harvesting, packing and handling should be done with great care to prevent damage to the highly sensitive, thin-skinned, turgid tubers. Crushing, Pressure Bruising, Brown Spot or Shatter Bruising are common defects and may lead to rapid water loss, shriveling and decay.

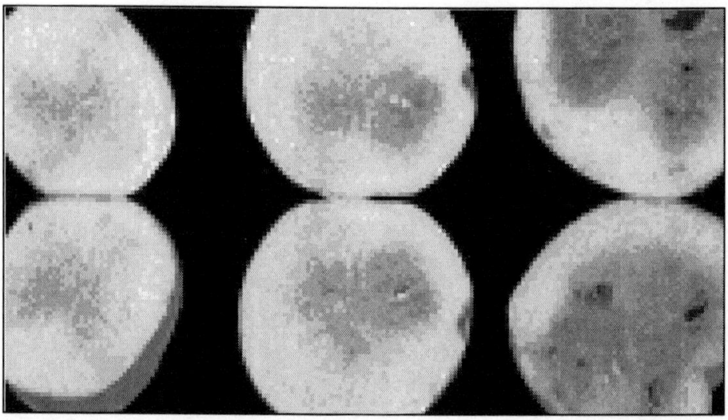

Chilling Injury

Air 0.2% **O_2 0.02%** **O_2**

Low Oxygen Effects

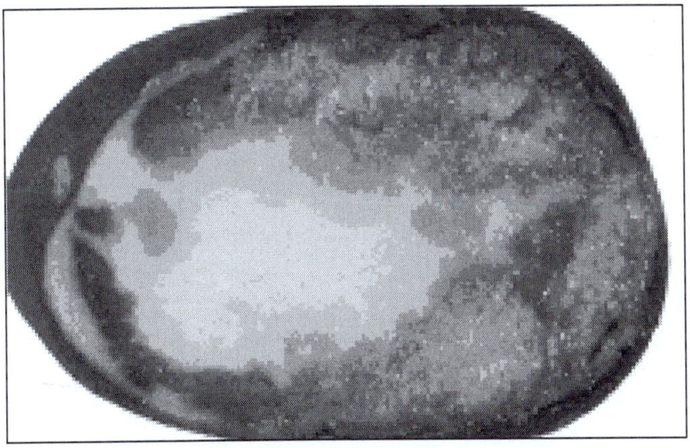

Bacterial Soft Rot

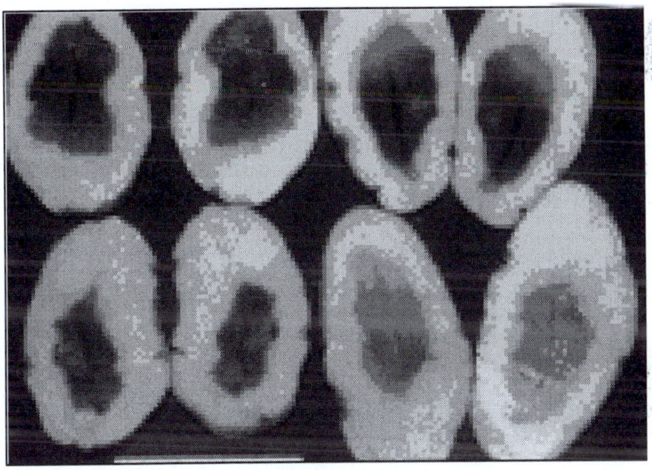

40°C Control

Black Heart

Brown Spot

Discoloration just beneath the inner surface of the tuber resulting from bruising or rough handling.

Freezing Injury

Freezing injury will be initiated at -0.8°C. Symptoms of freezing injury include a water-soaked appearance, glossiness, and tissue breakdown on thawing. Mild freezing may also result in chilling injury.

Pathological Disorder

Diseases are an important source of postharvest loss, particularly in

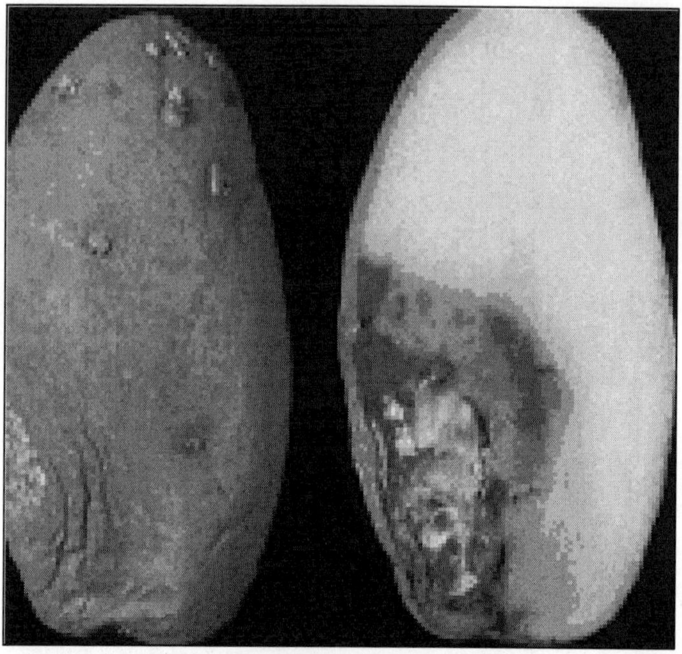

Fusarium Dry Rot

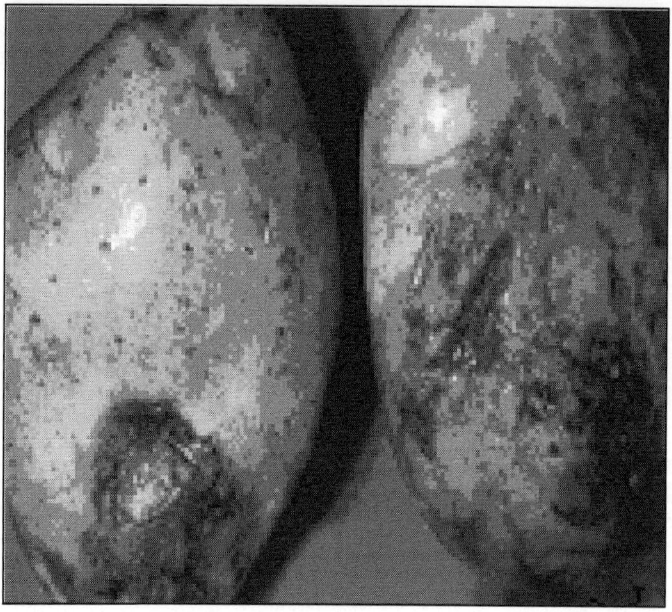

Late Blight

combination, with rough handling and poor temperature control. Three major bacterial diseases and a greater number of fungal pathogens are responsible for, occasionally, serious postharvest losses. The major bacterial and fungal pathogens that cause postharvest losses in transit, storage, and to the consumer are Bacterial Soft-Rot *(Erwinia carotovora* subsp. *carotovora* and subsp. *atroseptica), Ralstonia* (ex *Pseudomonas,* ex *Burkholderi) solanacearum,* Phytophthora infestans (late blight), Fusarium Rot (*Fusarium* spp.), Pink Rot (*Phytophthora* spp.), and Water rot (*Pythium* spp.) Occasionally serious diseases of immature tubers include Pink Eye (*Pseudomonas fluorescens*) and Grey Mold (*Botrytis cinerea*).

43

French Beans

Botanical Name: *Phaseolus volgaris*

Family: Leguminaceae

1. Introduction

It is an important leguminous vegetable, consumed as tender pods, and shelled green beans. A nutritious vegetable is largely grown in hilly areas of Himachal Pradesh, Jammu and Kashmir, and north eastern states during summer and winter and autumn crop in parts of U.P, Maharashtra, Karnataka. In northern plains, it is cultivated on autumn or spring crop, because of susceptibility to low as well as high temperature.

2. Composition (Per 100g Edible Portion)

Nutrient	*Composition*
Protein (g)	1.7 g/100g
Calcium (mg)	50 mg/100g
Phosphorus (mg)	28 mg/100g
Iron (mg)	1.7 mg/100g

Nutrient	Composition
Carotene (mg)	132 mg/100g
Thiamine (mg)	0.08 mg/100g
Riboflavin (mg)	0.06 mg/100g
Ascorbic Acid (mg)	24 mg/100g

3. Suitable Varieties

Arka komal, Bountiful, Contender, Jampa, Kentuky Wonder, Lakshami, Pant Anupma, Premier, Pusa Parvati.

4. Maturity Indices

Beans are harvested when they are rapidly growing and developing. Harvest occurs about 8-10 days after flowering for typical mature French beans. Beans should be harvested when the fruit is bright green, the pod is fleshy and seeds are small and green. After that period, seed development reduces quality and the pod becomes pithy and tough and looses green color.

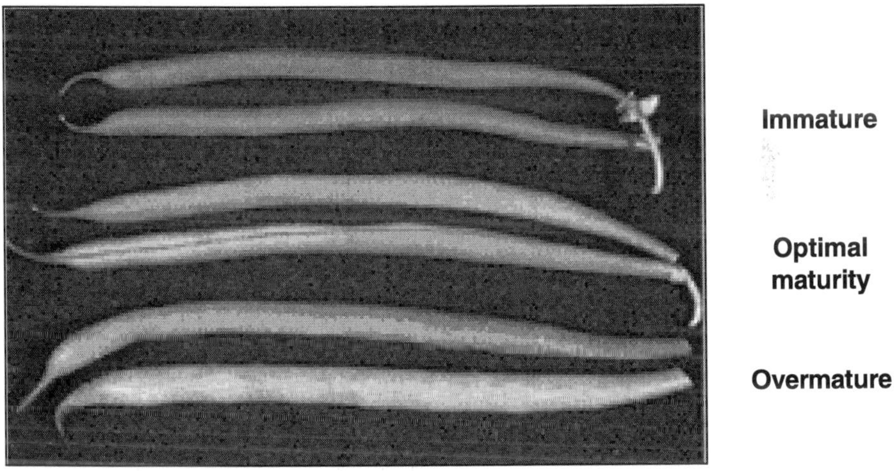

5. Quality Indices

Beans should be well formed and straight, bright in color with a fresh appearance, and tender but firm. They should snap easily when bent. Leaves, stems, broken beans, blossom remains, insect damage should not be present. Decreased quality during postharvest handling is most often associated with water loss, chilling injury, and decay.

6. Postharvest Management

Harvesting

French beans are harvested when pods they are tender (approximately 7-12 days after flowering). Bush beans mature relatively in a short period of 50 days requiring 2-3 pickings, while pole beans take 60-75 days and requiring 3-5 pickings. The beans are generally harvested by hand. Delayed harvesting reduces the quality of the pods as they become fibrous. Mechanical pickers are also devised for harvesting of beans especially for processing type varieties. These emply 'once over' destructive harvest, which strips leaves and removes pods from the plants..

Storage

(*a*) Optimum Temperature and Relative Humidity (RH)

5-7.5°C and 95-100 per cent (RH)

Very good quality can be maintained for a few days at temperatures below 5°C but chilling injury will be induced. Some chilling may occur even at the recommended storage temperature of 5°C after 7-8 days. At 5-7.5°C a shelf-life of 8-12 days is expected.

Water loss is a common postharvest problem with green beans. About 5 per cent weight loss is needed before shrivel and limpness is observed. After 10-12 per cent weight loss, the beans are no longer marketable.

(*b*) Rates of Respiration/CO_2 Production

Temperature °C	*ml CO_2>/kg hr*
0	10
5	17
10	29
15	46
20	65

To calculate heat production, multiply ml CO_2/kg hr by by 122 to get kcal/metric ton× day

(*c*) Rates of Ethylene Production

<0.05 µL/kg·hr at 5°C

(*d*) Responses to Ethylene

Exposure to ethylene at usual storage temperatures causes loss of green pigment and increased browning. Concentrations above 0.1 ppm reduce green bean shelf-life by 30-50 per cent at 5°C.

(*e*) Responses to Controlled Atmospheres (CA)

At recommended storage temperature, O_2 concentrations of 2-5 per cent reduce respiration rates. Snap beans tolerate and are benefited by CO_2 concentrations between 3-10 per cent. The main benefit is retention of color and reduced discoloration on damaged beans. Higher CO_2 (20-30 per cent) concentrations can be used for short periods, but can cause off-flavors.

7. Postharvest Disorders, Diseases and their Control

Physiological Disorders

Chilling Injury

The typical symptom of chilling injury in beans stored <5°C for longer than 5-6 days is a general opaque discoloration of the entire bean. A less common symptom is pitting on the surface. The most common symptom of chilling injury is the appearance of discrete rusty brown spots which occur in the temperature range of 5-7.5°C. These lesions are very susceptible to attack by common fungal pathogens. Beans can be held about 2 days at 1°C, 4 days at 2.5°C, or 8-10 days at 5°C before chilling symptoms appear.

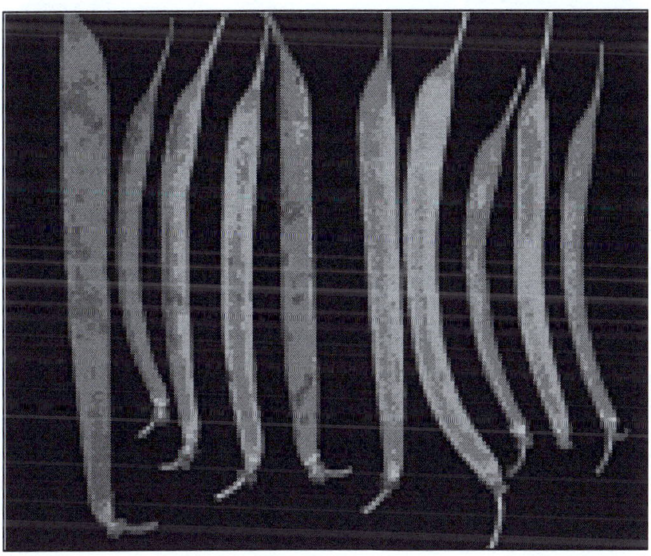

Chilling Injury Control

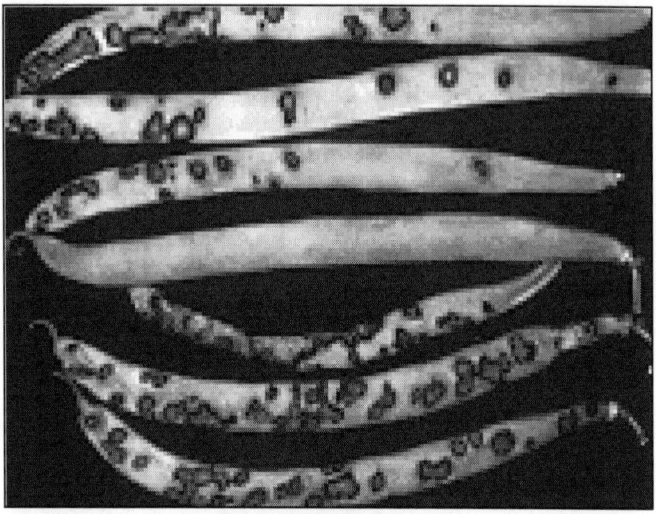

Anthracnose

Botrytis Gray Mold

White Rot

No discoloration occurs on beans stored at 10°C. Different varieties differ significantly in their susceptibility to chilling injury.

Freezing Injury

Appears an as water-soaked area, which subsequently deteriorate and decay. Freezing injury occurs at temperatures of -0.7°C or below.

Pathological Disorders

Decay due to various pathogens occurs after beans have been chill damage. Surface decay may also occur on stems and beans if free moisture is present during storage at >7.5. Common postharvest decay organisms on green beans are the fungi *Pythium, Rhizopus,* and *Sclerotinia,* all of which may occur as "nests" of decay or on broken or damaged beans.

Annexure 1

R&D Innovations in the Field of Postharvest Management

The post harvest technologies for loss reduction of Horticultural Commodities developed at CIPHET, Ludhiana and AICRP (PHT) centers

Technology Developed	*Application/Use*	*Description of the Technology*
Fruit and vegetable washing machine	To clean the fruit from dirt, soil, insect excreta and sap	A stainless steel portable mechanical washing machine (power 1 hp, cost Rs 25000-50,000 for 100-600 kg/h capacity) has been developed and commercialized by Ludhiana centre, suitable for a wide range of fruits and vegetables (carrot, potato, radish, turnip, ginger, okra, tomato, spinach, kinnow and pears). This machine could also be successfully used for breaking garlic bulbs and peeling potatoes.
Mobile cool chamber for transportation of fruits and vegetables	To transport the harvested fruits and vegetables to pre cooling chamber	Developed for short duration transportation of fruits and vegetables. The insulated box was designed such that it could hold 8 plastic crates of size 540x360x295 mm in two layer of four each for keeping fish. The total capacity of storage was 100 kg of vegetables with 80% filling of each plastic crates with ice trays the bottom. The cost of mobile cool chamber of is around Rs.20,000/- at the research fabrication.

Contd...

Technology Developed	*Application/Use*	*Description of the Technology*
Evaporatively cooled room for storage of fruits and vegetables	For short term cool storage of perishable commodities	An evaporatively cooled (EC) room (3x3x3m. inside size) was developed for on-farm storage of fruits and vegetables. The capacity of the unit is 2 tonne. The maximum temperature inside the EC room was 5-8'C lower than that inside the ordinary room and the minimum temperature was 5-8'C higher than that inside the ordinary room. Compared on the basis of 10% physiological loss in weight (PLW) the shelf life inside the room was 34 days for early kinnow, 23 days for late kinnow, 11 days fro cauliflower and 4 days for spinach. It was 21, 11, 5 and 2 days respectively for these fruits/ vegetables stored in an ordinary room at the same time.
CIPHET Evaporative Cooled Storage Structure	For short term storage of perishables	This system requires no or very minimum level of consumption of electricity, less initial investment and negligible maintenance cost. The structure has special design of roof, orientation and uses wetted pad as cooling medium can achieve 20°C lower temperature than outside temperature and relative humidity as high as 70–99 % depending upon the outside temperature. An ECS of about 5–7 tonne storage capacity may cost about Rs. 1.5 –1.8 lakh. It can be constructed at any place in farm preferable below the tree for better cooling efficiency.
Two stage evaporative cooler	To blow cool air in the storage shed or pack house	Developed two-stage cooler is portable and 1.5m x 1.0m x 2.0m in length, breadth and height. It could be able to drop the temperature up to the wet bulb depression and to 90 % relative humidity. The effectiveness of the two stage evaporative cooler ranged from 1.1 to 1.2 over the single evaporation. The hourly cooling capacity of TSEC ranged from 2125 to 4500 W.
Non-destructive measurement of maturity and sweetness of mango	For objective determination of the fruit quality on the tree	A color chart was developed to predict maturity of mango, which is simple and can be used on farm with the colorimeter. This nondestructive technique can be employed to sort the mango based on either maturity index or TSS (Sweetness) at export port, big mandies and in processing plant. It will also help in fixation of price of individual mango based on total soluble solids

Contd...

Technology Developed	Application/Use	Description of the Technology
Pomegranate Aril Extractor	To remove arils from fruit without damage	In the absence of proper hand tool we lose about 10% aril as food is first cut and then arils removed. To avoid the loss of arils and also spilling of juice while aril extraction, a simple hand tool has been developed at CIPHET. The counter rotation of the holders, breaks open the fruit into two irregular halves loosening the arils for easy separation. About 20-25% arils are already separated in the process of irregular breaking due to shearing action on the inner sheath and outer peel and rest are separated by hand.
Fruit Saving Gadget	Safe harvesting of fruits	Due to even short height throw of the fruit after plucking causes internal injury to the fruit and that leads to losses during transport and storage. To avoid damage to the fruit this simple device was developed at CIPHET. After plucking from tree, fruit is thrown in the trough of fruit saver and collected in a box/container. The fruits also get graded on size basis at appropriate places. Smallest grade fruit is collected first & largest grade is collected, in last."
Low Cost Basket Centrifuge for Minimal processing	To remove surface moisture from washed vegetables	Accumulation of surface water on the fresh and minimally processed vegetables is a matter of concern as it helps in growth of pathogens and microorganisms. A Basket centrifuge consisting of a detachable perforated cylinder, rotating at 500 RPM was fabricated to remove surface water from the minimally processed vegetables. The centrifuge was tested with minimally processed spinach and 'mustard leaves: "IT was bbsefveo Tlhat for minimally processed opinach and mustard, no leave injury was observable till 10 s. whereas, centrifugation of these vegetables beyond a period of 10 s. led to surface Injury.
Banana Comb Cutter	To safely remove banana combs	The banana bunches are transported green (unripe) from Bhusawal to rest of the country. Due to lack of proper tool to remove bunches for ripening, about 10-15% losses are observed in the form of cut bananas and also due to their un-uniform ripening. Hence a simple to use tool to de-stem the comb from the banana bunch has been developed at CIPHET. It replaces the use of sickle, which was much labor and time consuming. It is safe for human and banana hands and can be used for faster work

Contd...

Technology Developed	*Application/Use*	*Description of the Technology*
Dehumidified air dryer	To dry the green leafy plants at low temperature	A prototype dehumidified air dryer based on heat pump principle (cost Rs. 100,000 and capacity 25 kg/batch) has been developed by Bhubaneswar centre. The dryer suitable for drying of high value fruits, vegetables, spices and medicinal plants under low temperature and low humidity conditions to maintain excellent quality.

Literature

Bhutani, R C. 2003. *Fruit and Vegetable Preservation*. Biotech Books, New Delhi.

Bose, T.K, M.G. Som and J.Kabir. 1993. *Vegetable Crops*. Naya Prakash, Calcutta.

Bose, T.K, S.K. Mitra and D. Sanyal. 2002. *Fruits: Tropical and Sub Tropical*, Vol. 1. Naya Udyog. Calcutta.

Bose, T.K, S.K. Mitra and D. Sanyal. 2002. *Fruits: Tropical and Sub Tropical*, Vol. 2. Naya Udyog. Calcutta.

Canovas, Gustavo V Barbosa *et al.*, 2007. *Handling and Preservation of Fruits and Vegetables by Combined Methods for Rural Areas. Technical Manual/ FAO*. Daya Publishing House, New Delhi.

Chadha, K.L. 2001. *Hand Book of Horticulture*. ICAR, New Delhi.

Chandra Atul. 1993. *Arid Fruit Research*. Scientific Publications, Jodhpur.

Choudhury, Monalisa & Nayan Barua. 2006. *Marketing of Processed Fruit and Vegetable*. Daya Publishing House, New Delhi.

Cotton, Richard T. 2007. *Insect Pests of Stored Grain and Grain Products*. Biotech Books, New Delhi.

FAO. 2007. *Prevention of Postharvest Food Losses: Fruits Vegetables and Root Crops*. Daya Publishing House, New Delhi.

Fellows, Peter. 2007. *Guidelines for Small Scale Fruit and Vegetable Processors/ FAO*. Daya Publishing House, New Delhi.

Gangawane, L.V. & V.C. Khilare eds. 2008. *Crop Diseases: Identification and Management: A Colour Handbook*. Daya Publishing House, New Delhi.

Goel, Ashwani Kumar *et al.* eds. 2007. *Postharvest Management and Value Addition*. Daya Publishing House, New Delhi.

Gopalan, C., B.V. Rama Sastri and S.C. Balasubramanian. 2002. Nutritive Value of Indian Foods. (Revised). National Institute of Nutrition, ICMR. Hyderabad.

John, P Jacob. 2008. *A Handbook on Postharvest Management of Fruits and Vegetables*. Daya Publishing House, New Delhi.

Khetarpaul, Neelam. 2005. *Food Processing and Preservation*. Daya Publishing House, New Delhi.

Mazumdar, Bibhas Chandra & K Majumder. 2003. *Methods on Physico Chemical Analysis of Fruits*. Daya Publishing House, New Delhi.

Mazumdar, Bibhas Chandra. 2004. *Minor Fruit Crops of India: Tropical and Subtropical*. Daya Publishing House, New Delhi.

Morris, Thomas Norman. 2008. *Principles of Fruit Preservation 3rd Revised edn*. Biotech Books, New Delhi.

Salunkhe, D.K and S.S. Kadam. 1995. *Hand Book of Fruit Science and Technology: Production, Composition, Storage, and Processing*. Marcel Dekker, Inc. New York.

Salunkhe, D.K and S.S. Kadam. 1995. *Hand Book of Vegetable Science and Technology: Production, Composition, Storage, and Processing*. Marcel Dekker, Inc. New York.

Index

p. 9

SO_2 Injury p.13

Grey Mould p.15

p.16

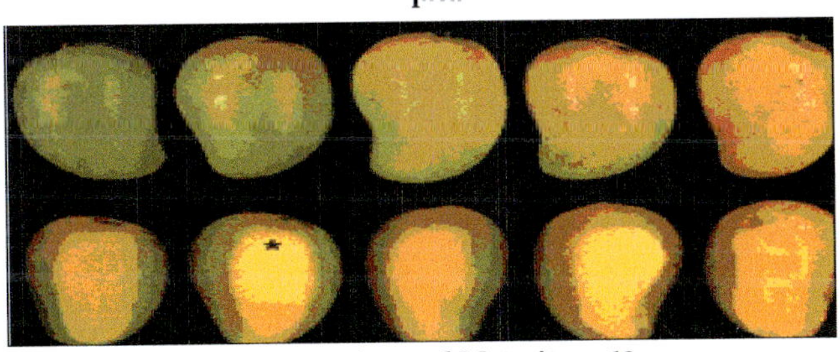

Fruit Ripening and Maturity p.19

Cold Storage/Low Temperature Injury p. 22

Anthracnose p. 22

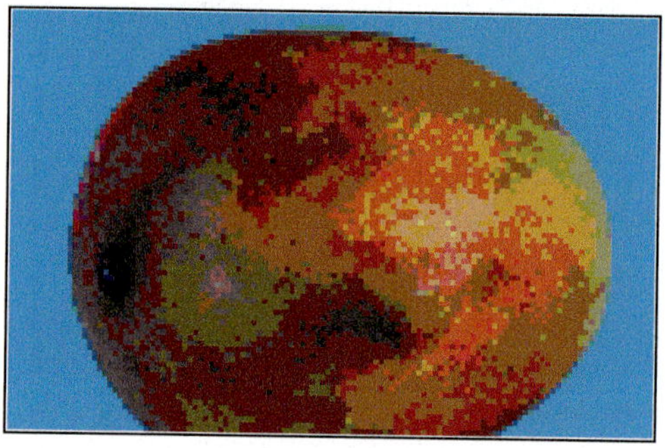

Stem End Rot Jelly p. 23

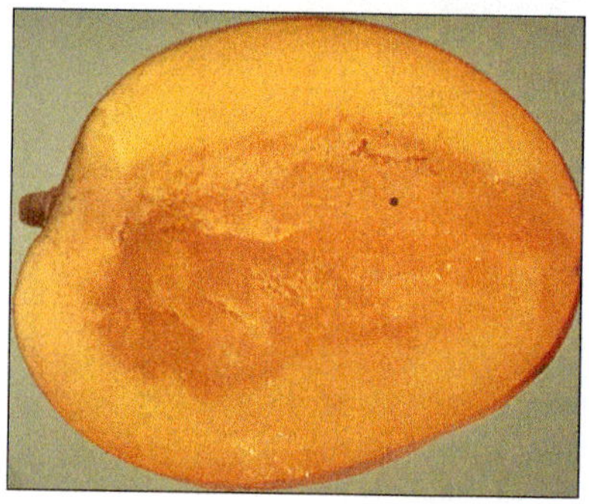

Jelly Seed p. 23

p. 24

Brown Spot p. 24

1 2 3 4 5 6 7

Different Stages of Maturity p. 26

Water Blister p. 29

Fruitlet Core Rot p. 29

Water Blister p. 30

Maturity Stages p. 32

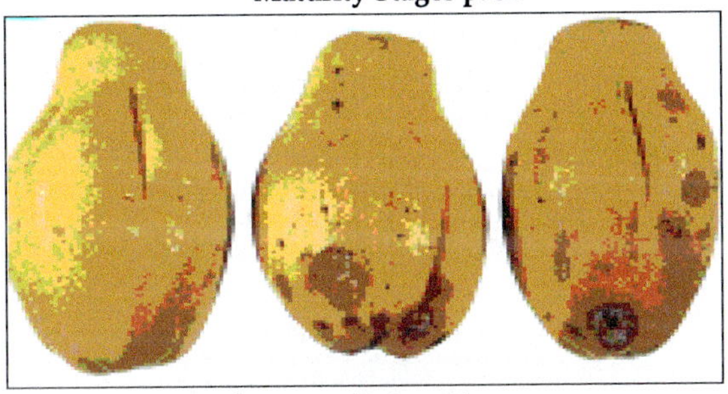

Anthracnose Rot p. 36

Alternaria Rot p. 36

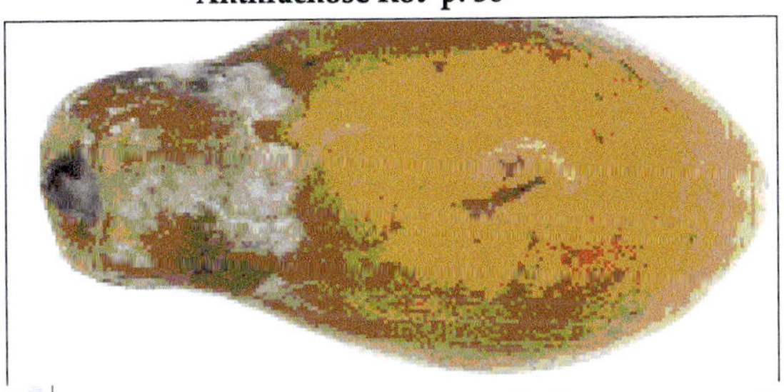

Phytopthora Stem End Rot p. 36

p. 38

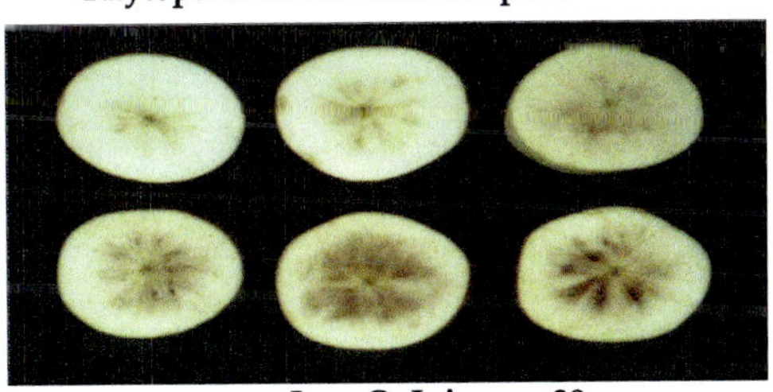

Low O_2 Injury p. 38

Blue Mould p. 38

Grey Mould p. 38

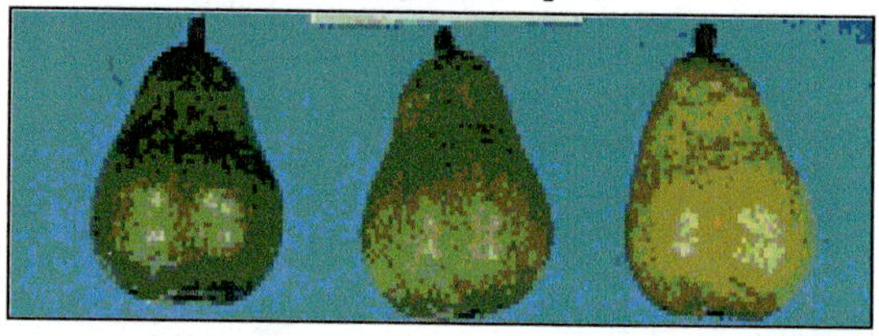

0 hr **6 day** **2 day**

1 day **12 hr** **Ambient 6 day**

p. 38

p. 45

Abrasion Damage p. 52

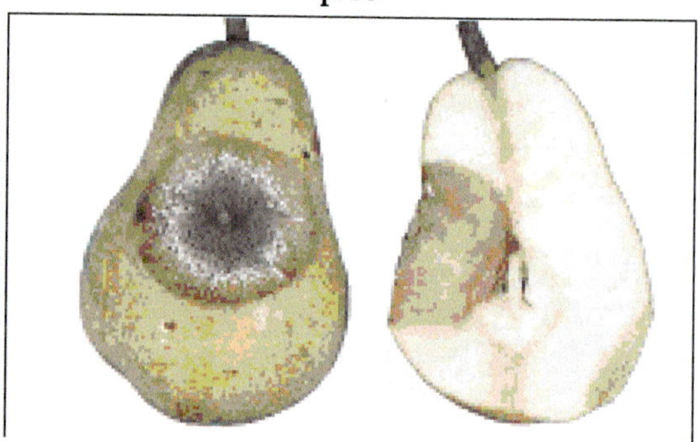

Blue Mould Rot p. 52

Air **0.25% O_2** **80% CO_2** **0.25% + O_2 80% CO_2**

After 3 Days at 20°C p. 53

Freezing Damage of Pear p. 53

p. 54

Ripeness Stages p. 55

p. 61

Wounding during Harvest and Handling p. 66

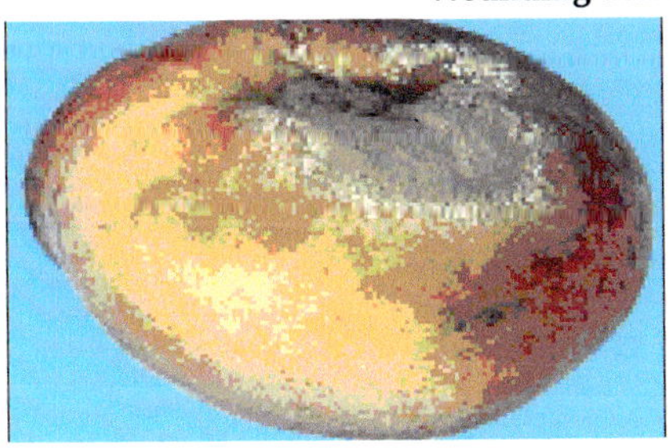

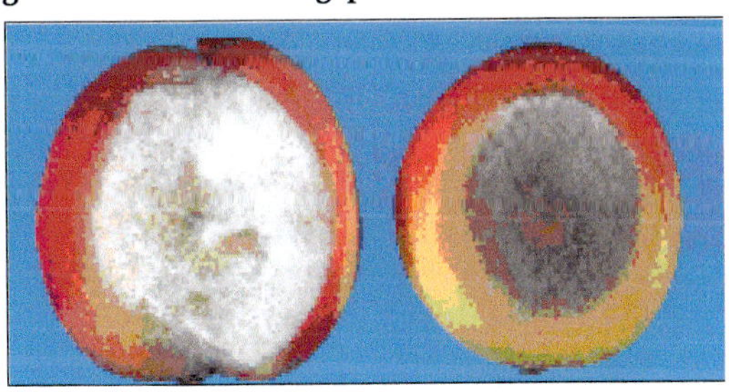

Grey Mould p. 66

p. 67

3.5 Month Storage p. 67

8 weeks at 10ºC+ 3 days at 20oC

8 weeks at 2.2ºC + 3 days at 20oC

Chilling Injury p. 72

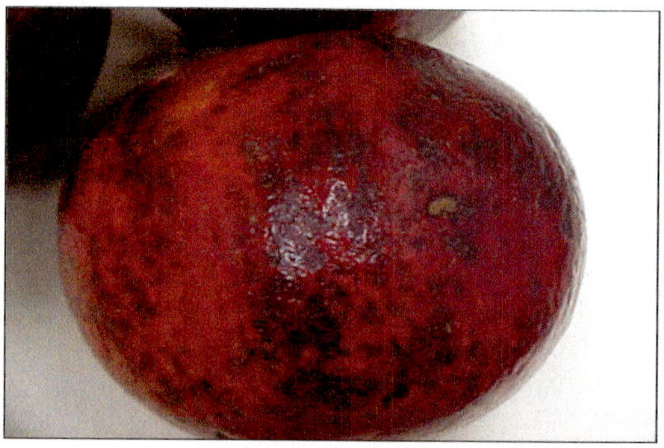

Sun Scald p. 72

Sunburn p. 72

p. 74

Maturity and Ripeness Stages p. 74

Brown Rot p. 78

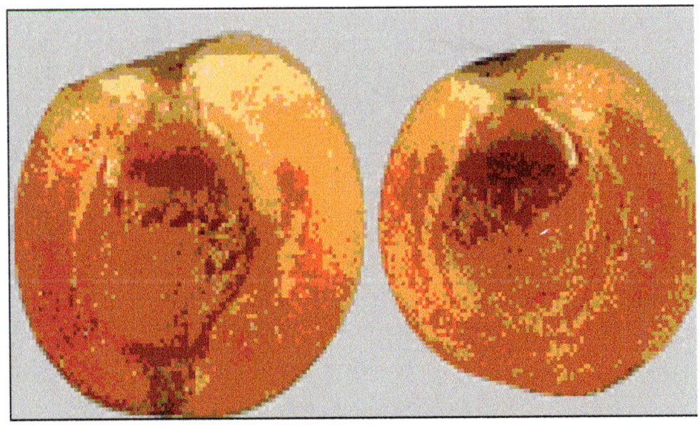

Pit Burn p. 78

Mechanical Damage p. 79

Mechanical Damage p. 80

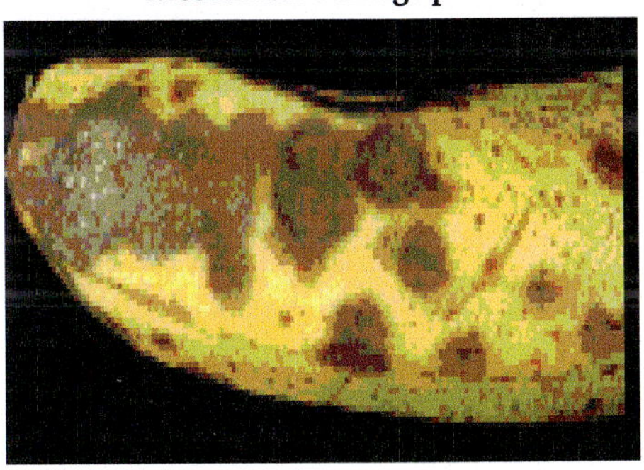

Anthracnose p. 84

Cigar End Rot p. 85

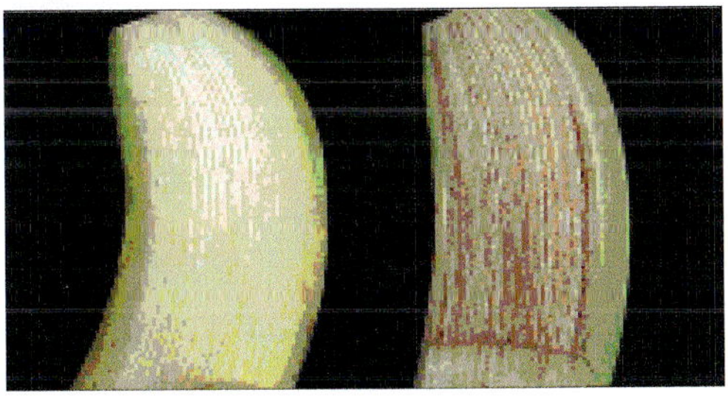

Control Chilling Injury p. 84

Crown Rot Chilling Injury p. 85

Chilling Injury p. 86

p. 87

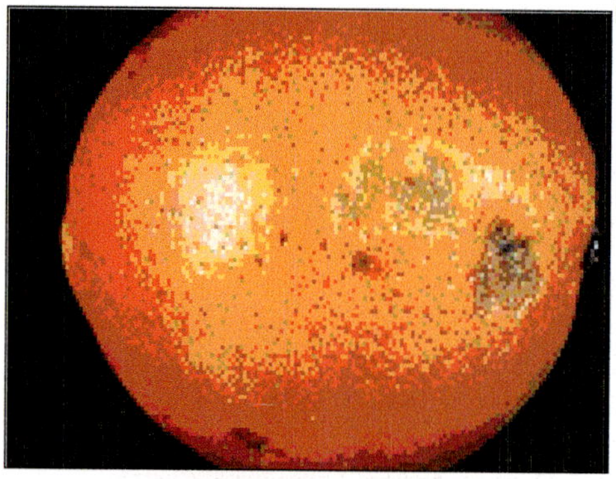

Chilling Injury p. 95

p. 92

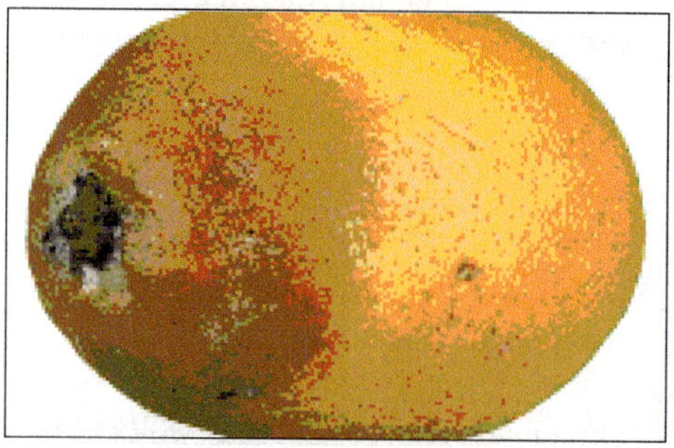

Anthracnose p. 95

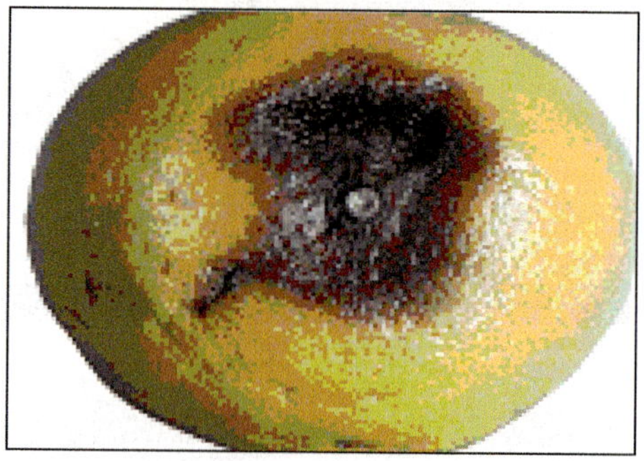

Alternaria Rot p. 96

B. Cinera Botrytis Rot p. 96

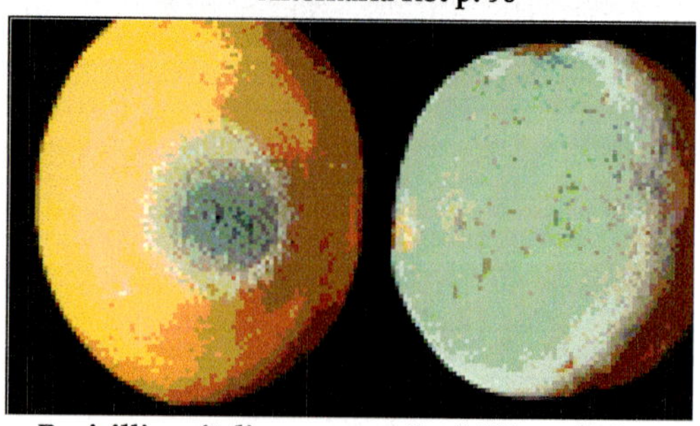

Penicillium italicum **Penicillium digitatum**

Blue Green Rot p. 96

Phomopsis citri Phomopsis p. 97

p. 103

p. 9[illegible]

Concentric Cracking p. 112

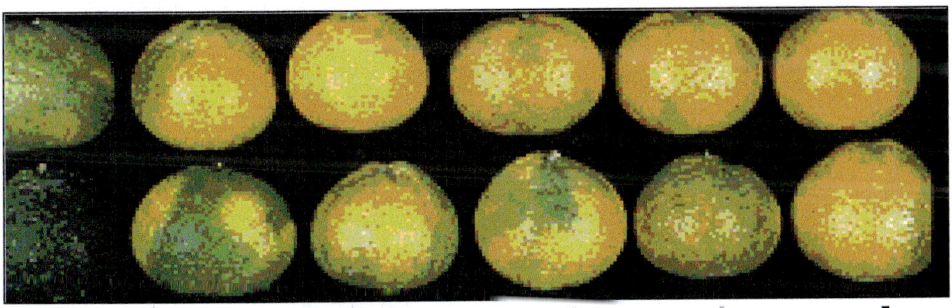

0 1 2 3 4 5

Degreening p. 102

p. 108

Cherry Maturity Stages p. 110

Cherry CA Effects p. 112

p. 115

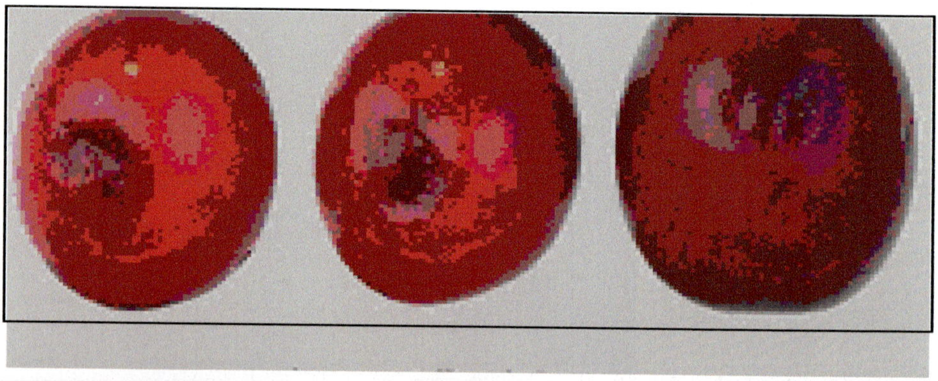

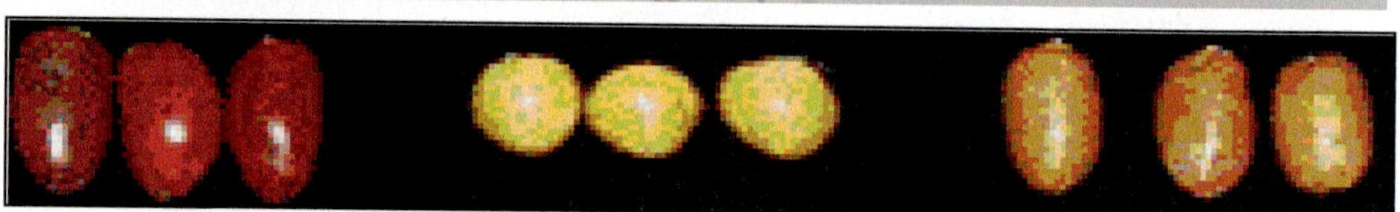

Abada Barhee Deglet Noor

p. 116

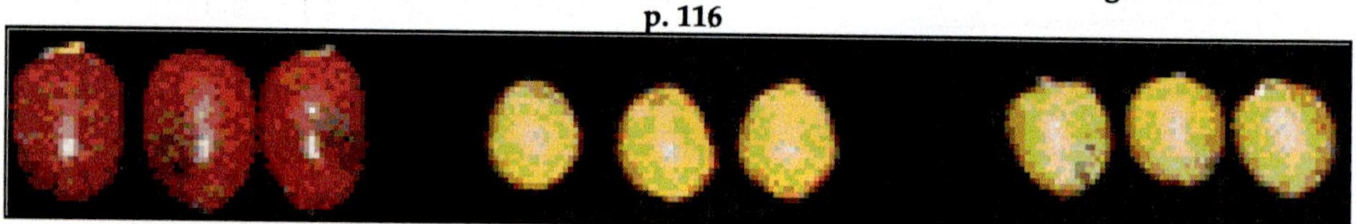

Halway Honey Khadrawy

p. 116

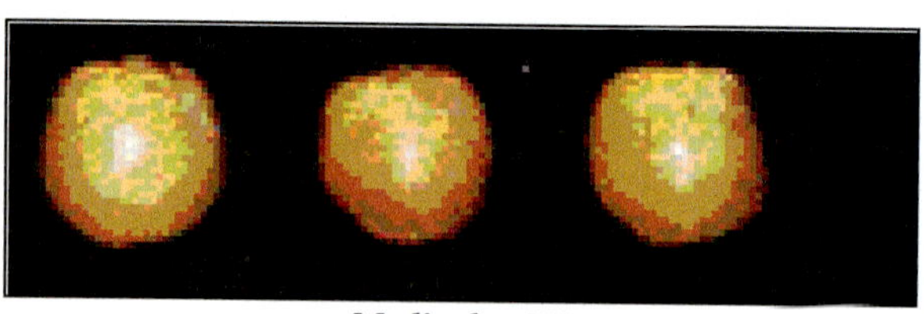

Medjool p. 116

p. 121

Khalal Partially-rutab Rutab
p. 117

p. 128

Quality p. 118

Air + 20 per cent CO_2 After 3 weeks 5ºC

CO_2 Effect on Calimyrna p. 118

p. 126

Fruitlet Quality p. 128

p. 130

p. 137

Quality Fruit p. 139

Sun Burning p. 141

Fruit Cracking p. 141

p. 143

p. 147

Ripeness Stages p. 149

p. 153

Control p. 156

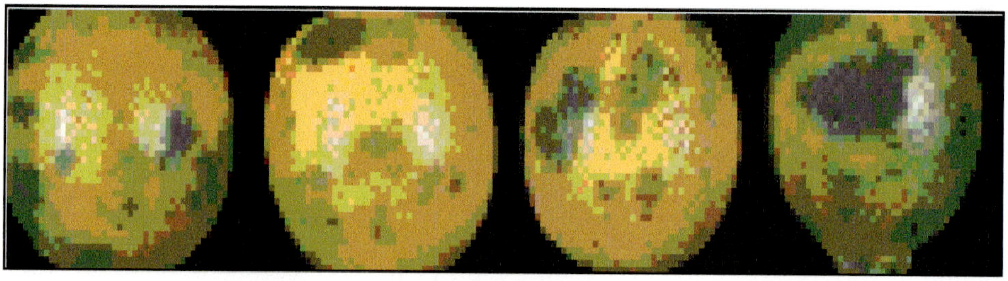

Oil Spotting p. 156

Oil Spotting p. 158

Anthracnose p. 161

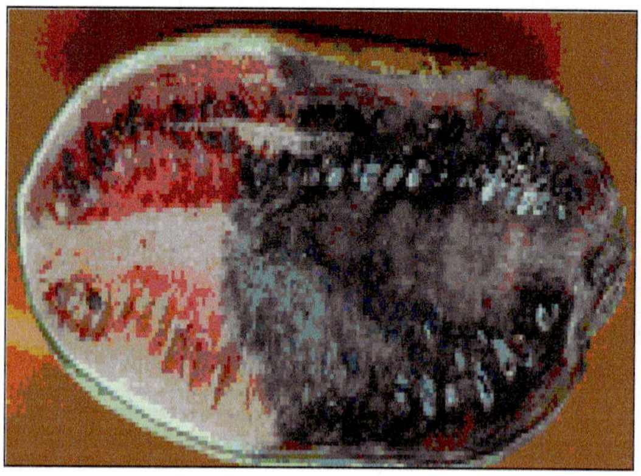

Lasiodiplodia Rot p. 162

p. 164

Botrytis Stem End Rot p. 162

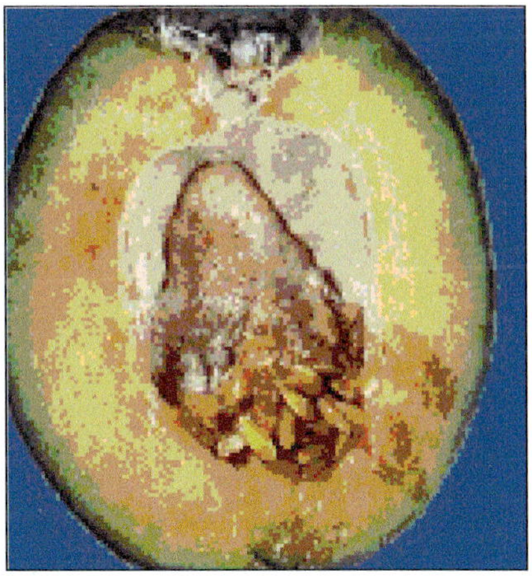

Sour Rot p. 168

Rhizopus Rot p. 168

p. 171

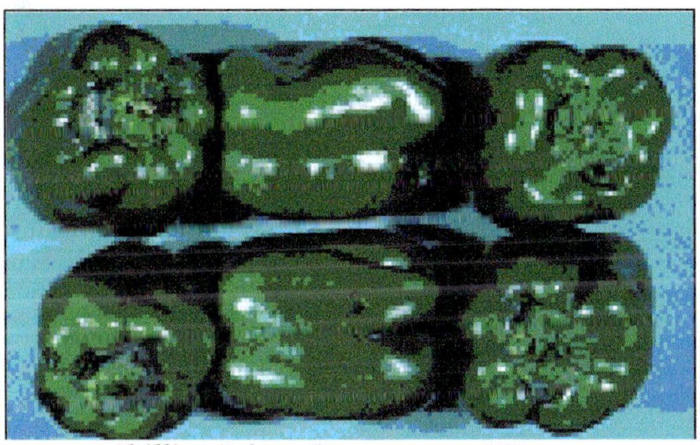

Chilling Injury Symptoms p. 175

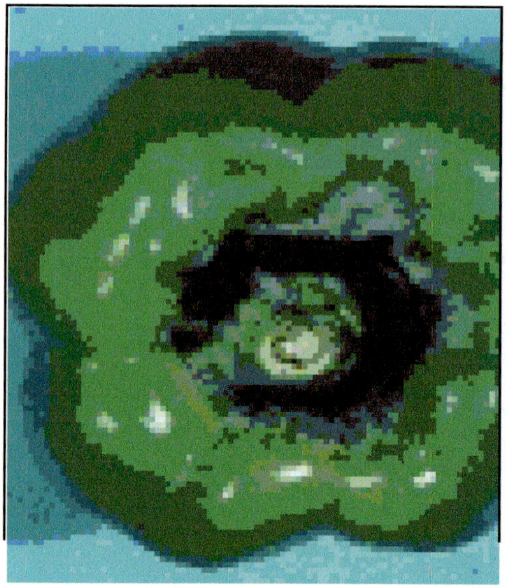

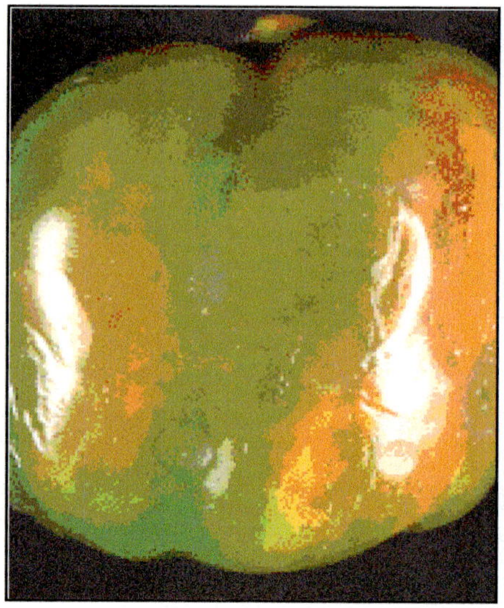

Bacterial Soft Rot p. 176

p. 177

p. 182

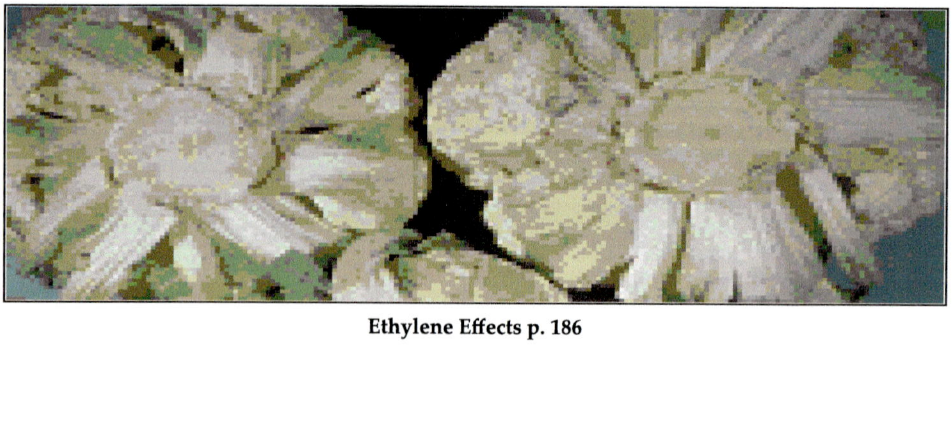

Ethylene Effects p. 186

(Physical damage followed by Alternaria spp.)
Mechanical Damage p.187

Solar Yellowing p.187

p.188

p.195

Bacterial Soft Rot Black Mold p.192

Chilling Injury p.198

Ethylene Damage p.199

Fusarium Decay p.199

p.201

Chilling Injury p.205

Solar Yellowing p.206

Sun Scald p.201

p.207

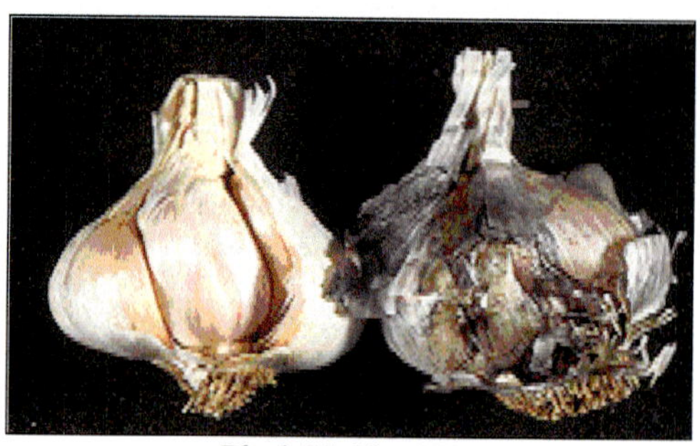

Black Mold p.211

Sprout Development p.211

p.213

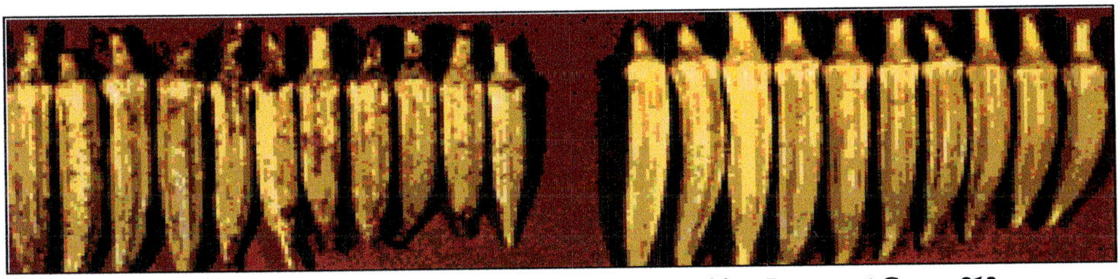

Air Control

Air + 5 per cent Co_2 p.213

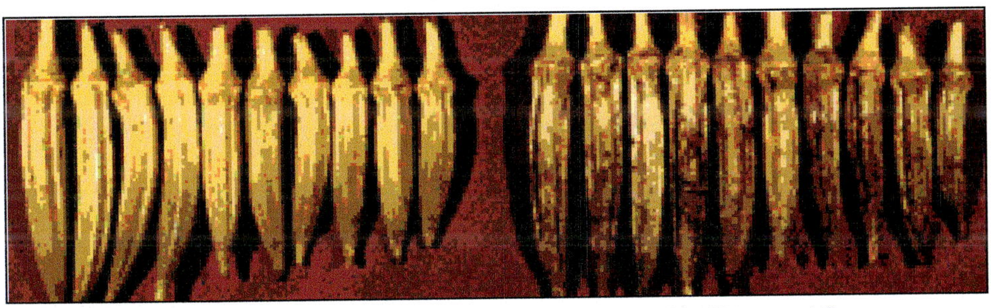

Air + 10 per cent Co_2

Air+15 per cent Co_2

Chilling Injury p.213

p.218

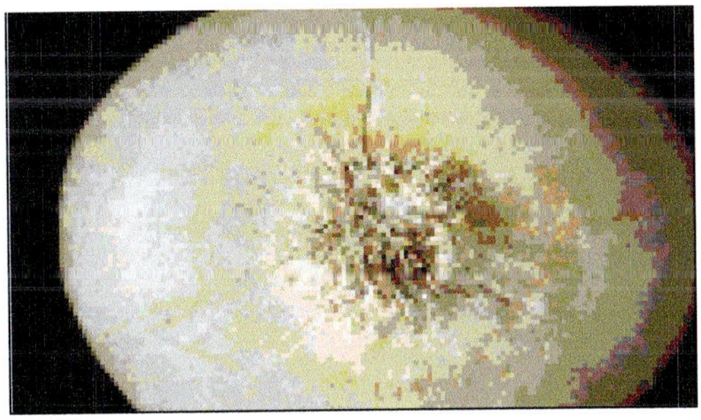

Fusarium Basal Rot p.222

Grey Mold p.223

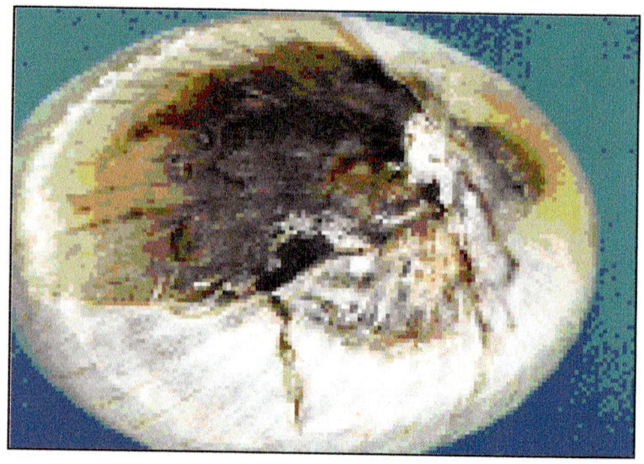

Fusarium Bulb Rot p.222

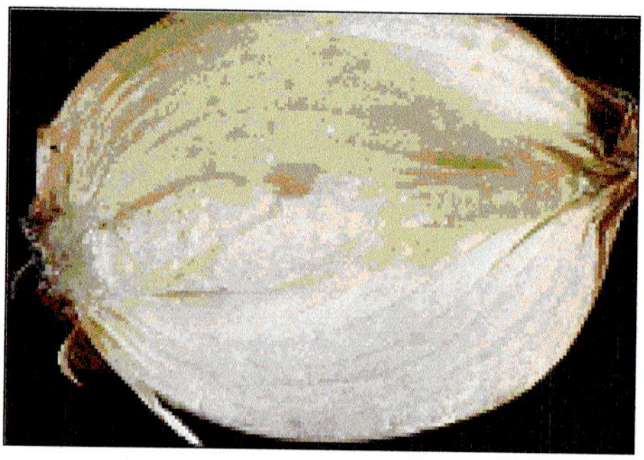

Soft Rot p.223

Translucent Scales p.223

p.231

p.225

Frost Damage p.229

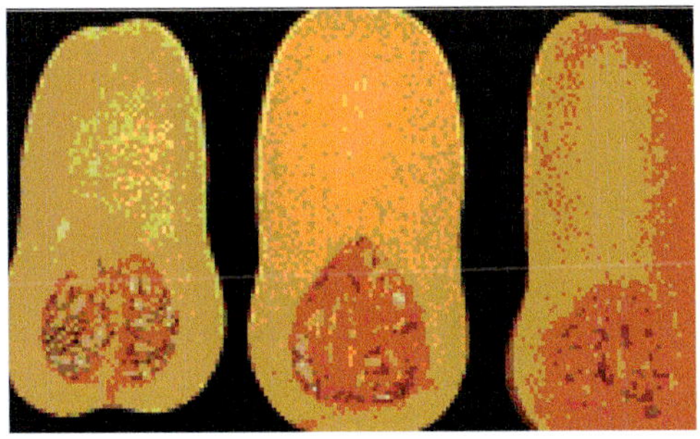

Butternut Squash Internal Color Scale p.234

p.236

p.240

p.244

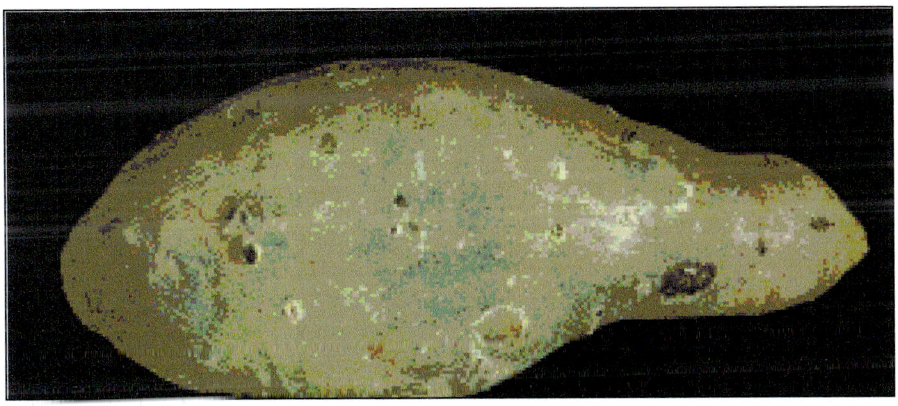

Botryodiplodia Rot p.240

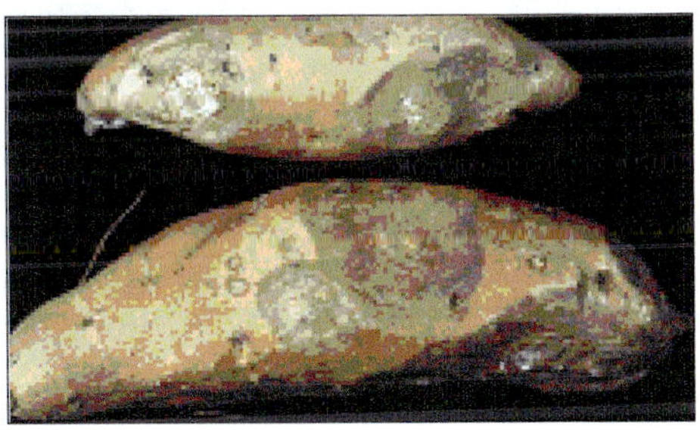

Fusarium Rot p.248

Rhizopus Rot p.248

p.250

Surface Abrasion p.249

MG: Mature-green Tomato Maturity Stages p.252

Cherry Tomato Maturity and Ripeness p.253

Effects of Temperature on Tomato Effects of High Temperatures on Tomatoes p.256

Control Ethephon

Ethephon Effects on Tomato p.257

p.261

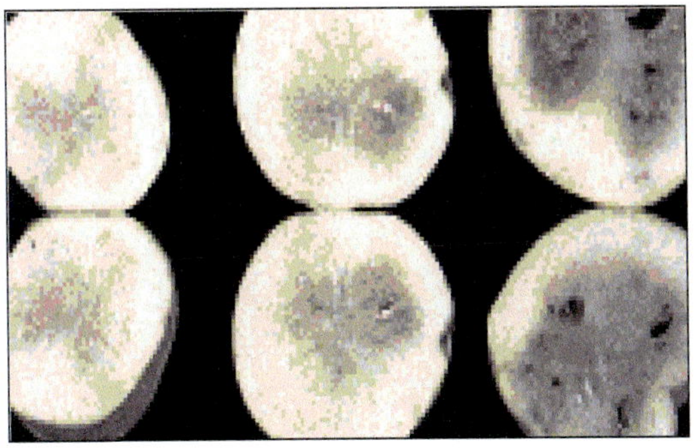

Chilling Injury p.266

Low Oxygen Effects p.266

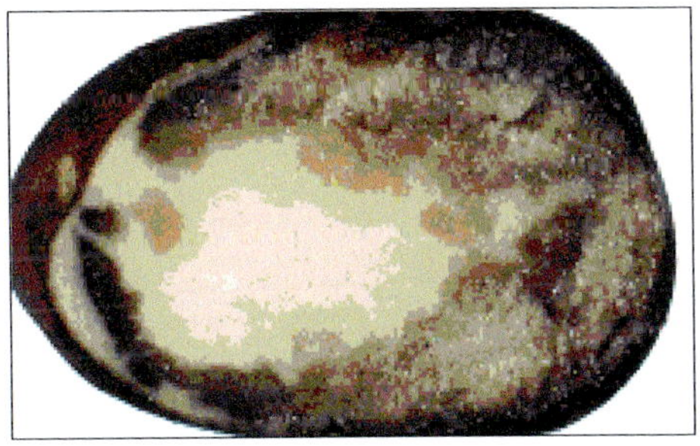

Bacterial Soft Rot p.267

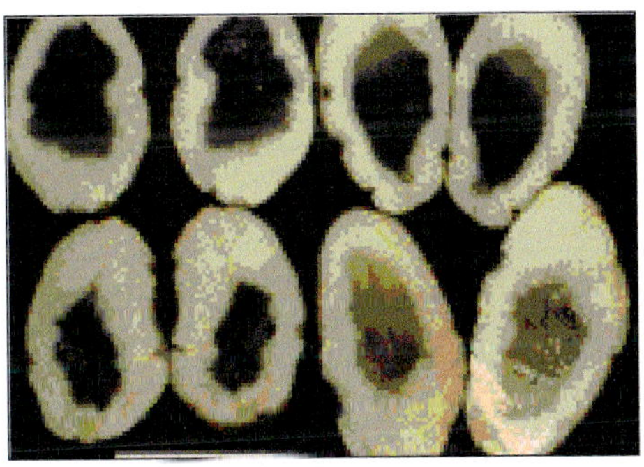

40ºC Control Black Heart p.267

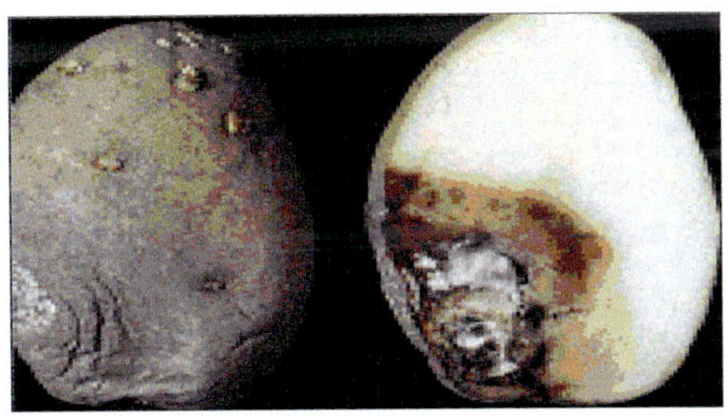

Fusarium Dry Rot p.267

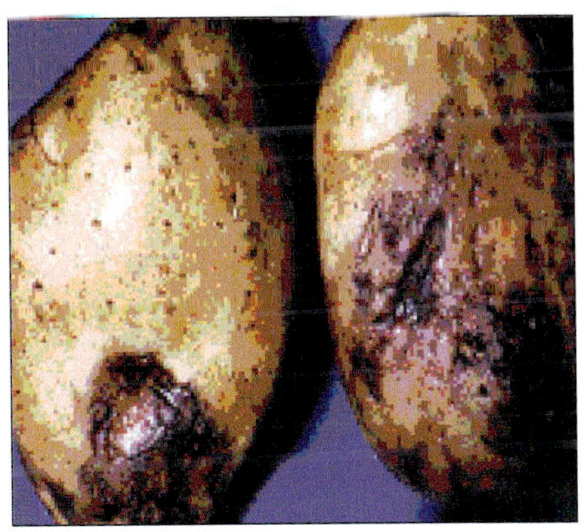

Late Blight p.268

p.270

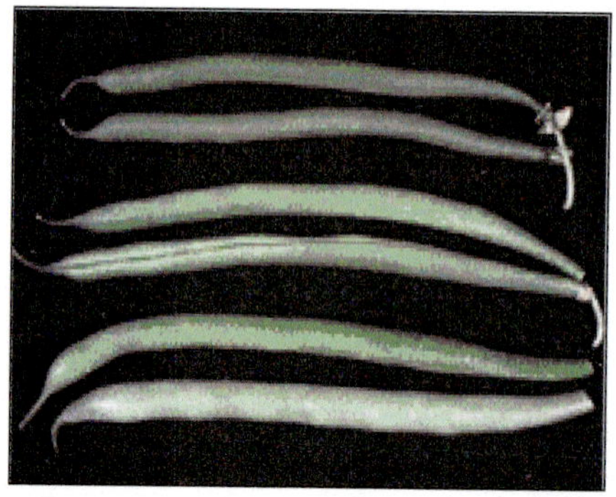

Immature Optimal maturity Overmature p.271

Chilling Injury Control p.273

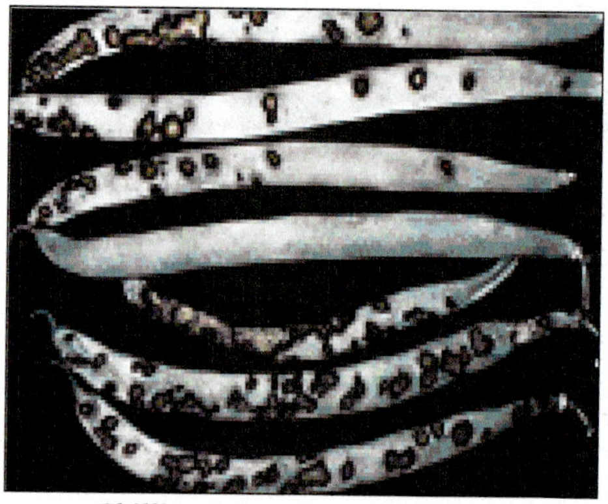

Chilling Injury Control p.274

Botrytis Gray Mold p.274

White Rot p.274